Roger L. Wabeke
MSc, MScChE, CIH, CHMM

AIR CONTAMINANTS
and
INDUSTRIAL HYGIENE VENTILATION

A Handbook of Practical Calculations, Problems, and Solutions

LEWIS PUBLISHERS

A CRC Press Company
Boca Raton   London   New York   Washington, D.C.

Library of Congress Cataloging-in-Publication Data

Wabeke, Roger L.
  Air contaminants and industrial hygiene ventilation: A handbook of practical calculations, problems, and solutions / Roger L. Wabeke.
      p. cm.
  Includes index.
  ISBN 0-56670-307-7
  1. Factories—Heating and ventilation—Handbooks, manuals etc. 2. Industrial hygiene—Handbooks, manuals etc. 3. Air—Pollution—Prevention—Handbooks, manuals etc. I. Title.
RC963.3.W32 1998
363.729'2—dc21   98-4620

Visit the CRC Press Web site at www.crcpress.com

© 1998 by CRC Press LLC
Lewis Publishers is an imprint of CRC Press LLC

No claim to original U.S. Government works
International Standard Book Number 0-56670-307-7
Library of Congress Card Number 98-4620
Printed in the United States of America   3 4 5 6 7 8 9 0
Printed on acid-free paper

Dedication

This book is dedicated to those who have endeavored, currently strive, and will push to improve the air quality of our workplaces, homes, and all atmospheric environments in our global community.

The Author

Roger L. Wabeke, board certified since 1973 in comprehensive industrial hygiene practice, has 33 years of professional experience. His primary interests are chemical safety engineering and occupational toxicology. His undergraduate education and training were in biology and chemistry. His graduate studies led to a Master of Science degree in Occupational and Environmental Health and a Master of Science degree in Chemical Engineering, both from Wayne State University, Detroit, MI. He is board certified (Master's level) in hazardous materials management.

Mr. Wabeke is an adjunct professor in the Departments of Occupational and Environmental Medicine at Wayne State University School of Medicine and in Pharmacy and Allied Health and Community Health at Wayne State. He has authored many publications in areas of his professional interests. Currently, he is a principal investigator in a major National Cancer Institute case-control research project studying the occupational epidemiology of prostate cancer. He is a member of several professional organizations and, over the years, has received many distinguished appointments and awards.

During his career, Mr. Wabeke evaluated and controlled exposures of many thousands of workers to chemical and physical agents in numerous, diverse workplaces. He consults for the chemical, pharmaceutical, petrochemical, and nuclear energy industries; the military; academic institutions; healthcare facilities; environmental pollution control projects; real estate and financial institutions; and airline, trucking, maritime shipping, and pipeline transportation companies, among others. He practiced industrial hygiene for five years for BASF Corp., a major chemical manufacturing company. He was an industrial hygienist for 16 years for Ford Motor Co., 10 of which he was supervisor of industrial hygiene for worldwide operations. Since 1987, he has been president of Chemical Risk Management, an environmental and occupational health consulting firm with offices and laboratory in Detroit and Dearborn, MI.

Author Disclaimer

Conscientious effort was made to ensure that the contents of this handbook and every problem, in particular, are technically accurate, complete, and useful in the day-to-day practice of contemporary industrial hygiene. All calculations and problems were critically peer reviewed. However, when many thousands of information items are entered into a published work, a few typographical errors can result even with the best effort of everyone involved in the process. To ensure completeness and accuracy of the calculations, users are encouraged to send corrections, additions, and comments which enhance the usefulness of this handbook to the author or to the publisher.

Roger L. Wabeke
Detroit, MI

Preface

Man is here for the sake of other men, and also for the countless unknown souls with whose fate we are connected by a bond of sympathy.

— Albert Einstein

Several problems and calculations in this handbook were developed initially to assist industrial hygienists preparing for their board certification examinations. After the author passed his board certification examinations in Boston in 1973, he and several other certified industrial hygienists organized a one-day, admittedly crude, review course for others who were preparing for their examinations. The course was offered semiannually, on Saturday, at Wayne State University College of Medicine in the Department of Occupational and Environmental Health. The Michigan Industrial Hygiene Society provided free refreshments. The registrants and instructors brought "brown bag" lunches. There were no fees for this course, and the instructors provided *pro bono* lectures. Nine discussion rubrics covered air sampling and analysis, ventilation, toxicology, calculations, radiation, respiratory protection, industrial hygiene chemistry, noise, and heat stress. With no more than 50 minutes for each subject, only industrial hygiene "pearls" could be presented. Today, of course, there are several high quality one-week review courses offered around the country for a substantial fee. The depth and scope of these courses are now far more encompassing and preparative.

Many of the problems in this book were acquired during the author's career from resolving actual industrial hygiene exposure issues including evaluations of numerous workplace poisonings and fatalities. A few were provided over the years by present and former colleagues, mentors, and students far too numerous to mention. Still others served as the basis for homework assignments and examination questions for students in classes which the author presented at Wayne State University School of Medicine, Henry Ford Community College, and the University of Michigan School of Public Health.

How To Use This Book

This handbook is intended primarily for use by:

- ✎ Industrial hygienists
- ✎ Heating, ventilation, and air conditioning system (HVAC) engineers
- ✎ Air pollution control engineers
- ✎ Chemical safety engineers
- ✎ Hazardous material managers
- ✎ Inhalation toxicologists
- ✎ Air contaminant emergency responders
- ✎ Environmental control engineers
- ✎ Atmospheric scientists
- ✎ Professors and teachers of these subjects
- ✎ Graduate and undergraduate students in these areas

Several typical plus some uncommon industrial hygiene problems which require mathematical solutions are covered. "Tips" are given to help one prepare for and take the board certification examinations. Common formulae, equations, conversions, and other information worthy of committing to memory and practice are also included.

This book is designed to be browsed. Those preparing for the board certification examinations should master the introductory chapter and, at a minimum, the following 14 problems: 140, 304–306, 308–314, 271, and 276–277. Once these are successfully handled, problems 1–7, 10–14, 16–20, 27, 29–30, 32, 70, 79, 88, 93, 101–102, 108, 297, 316, 321, and 406 should be mastered. Those in the best position to achieve a high score on these aspects in certification examinations will begin preparing at least one year in advance. If only two problems are mastered daily, five days a week, for 45 weeks during the preparatory year, every problem and exercise in this book will have been covered.

The teacher of these subjects can extract selected, relevant course problems for student assignments and homework. With only slight modifications, each problem can be custom "tailored" to make it unique with pedagogic relevance to the course materials and content. Many of the problems can serve as a launch point to discuss industrial hygiene control methods or the consequences to workers' health if preventive steps are not taken.

The seasoned, experienced professional is encouraged to browse through this book as well. Colleagues have suggested that solving these problems might be used to help maintain board-certification maintenenance points; however, this concept should be broached with the American Board of Industrial Hygiene. Moreover, the Board may select key problems from this casebook to be included in the revolving competitive certification examinations.

The problems in this workbook are not grouped into categories. Few problems in industrial hygiene fall neatly into a specified "type." Rather, most problems faced by industrial hygienists envelop several diverse topics. For example, a complex chemical spill issue could entail compound calculation sets of evaporation rates, air contaminant concentrations, worker's dose determinations, additive mixtures, community evacuation parameters, and dilution ventilation requirements. Several problems encompass such a broad scope and are of this nature (e.g., see problem 316, referred to by a colleague as the "mother of all industrial hygiene problems"). Some problems, understandably, are more rigorous than others. Certain key problem types are presented in various ways.

And, since the problems in this handbook cannot be easily sorted into groups, it is the author's hope that industrial hygienists, chemical safety engineers, students and their teachers, atmospheric toxicant scientists, ventilation engineers, and others will frequently browse through the problems and the index. Only in this way will the users begin to develop calculation methods and their "mental index" of the broad scope and diversity of the plethora of problems and their own systematic schema to define and solve them.

Another purpose of this handbook is to provide an assortment, a repository, and a reference set of calculations to assist industrial hygienists throughout their careers. Calculations to arrive at solutions to many of these types of problems are performed frequently by industrial hygiene practitioners. Others are done less often, some rarely, but the examples are included to assist resolution of uncommon problems. The author hopes that students of these problems will not only be challenged, but will also see the diversity of issues which industrial hygienists and chemical safety engineers encounter, and that they will be stirred to regularly return to this handbook and prowl through these problems. In so doing, those responsible for conserving the health and providing for the comfort of workers and the public's health and safety might see solutions applicable to problems which they regularly, or perhaps even rarely, encounter in their professional practice.

Many problems can be solved by a variety of methods. The calculations are not always necessarily the "best" or the quickest way; however, the author was comfortable with the approach that was taken and fully recognized that stated solution steps might not be the most expedient method to arrive at the answers. Most problems, hopefully, balance theory and practical applications. A tad of levity is injected into some problems to help break the tedium and chore of some of the more daunting calculations.

These problems do not require mathematical skills beyond college algebra and elementary statistics. Most calculations are simple arithmetic, but all require humility and critical, logical thinking based on sound understanding of basic industrial hygiene principles relating to evaluation of exposures to air contaminants. Since the science aspects of industrial hygiene (as contrasted with intuitive "art" parts) are quantitative in nature, those who are adept at number "crunching" and mathematical logic should have no difficulty solving these problems. Once the principles of a problem are understood, then it becomes, as the engineers say, simply a matter to methodically "plug and chug" the often ponderous numerical arithmetic parts. Several of the problems in this workbook are variations on a common theme. The author believes that, to fully understand some of the key concepts, one must be able to see a multi-faceted problem from all aspects and be able to solve for the different variables.

Finally, these problems are a work practice module. For those preparing for the board certification examinations, little will be gained by sitting down and just reading these problems and skipping the exercises. One will reap only nominal benefit from the problems unless they are systematically analyzed, comprehended, and completed.

Remember that the certification examination questions are highly quantitative in nature. Over half of the core aspects examination questions may involve questions that require calculations, whereas the comprehensive practice portion may be 20% or more questions requiring calculations in air pollution, noise, radiation, heat stress, chemistry, ventilation, statistics, toxicology, safety engineering, ergonomics, and other rubrics.

Since there is no set order of topics in industrial hygiene air and ventilation calculations, the *Index to Problems* may be used to find the types of problems you wish to master. There are numerous methods to solve most of these problems. The solution methods usually are related to each other, but may seem very different. In this workbook, some related problems are solved using one method, and some by another. You should attempt to do the problems yourself before looking at the given solutions. If you get the correct answer using a reasonable method, you need not worry that you did not use the method which was presented in this workbook. If the method presented is clearer than the one that you used, however, you might consider adopting it for similar problems in the future.

Contents

Introduction

AMERICAN BOARD OF INDUSTRIAL HYGIENE CERTIFICATION EXAMINATION HINTS

1. Bring a scientific calculator with fresh batteries. Be able to apply all major functions. The calculator should have scientific notation, $\log_{10}$ and natural logarithms, common conversions (e.g., gallons to liters, lb to kg, °F to °C, etc.), exponential notation, and basic statistical functions.
2. Bring sharp pencils and new ball point pens to the exams. You might consider hard candy, mints, and gum (nonbubble type). How about a canteen of cold juice? Or a Thermos® of coffee? Cans of caffeineated soda?
3. Do not "cram" on the nights before the examinations.
4. Get a good night's sleep. Arrive refreshed and confident.
5. Wear comfortable clothes, e.g., big, soft, oversized shoes are nice at examinations. Sit in the center of the examination room to avoid any cold drafty walls and windows, solar heat, excessive glare, and contrasty shadows. Select a comfortable chair.
6. Eat a light, well-balanced, nourishing breakfast and a similar lunch. Avoid heavy pancakes, fatty food, highly fibrous food, greasy doughnuts, etc. Sugars from fresh fruits and a couple sources of proteins might not be a bad idea. Thinking brains need amino acids. Bring slices of processed cheese to the examinations. Studies show we reason better when we are well hydrated. Taking a laxative the night before is not prudent. Be mindful that the residence time of food in the gastrointestinal tract can be 24 or more hours.
7. **Your first examination calculation should be:**

$$\frac{\text{examination duration (in minutes)}}{\text{total number of questions}} = \frac{\text{average number of minutes allocated}}{\text{question}}$$

   Be mindful of this average, and do not spend too much time on any single question. Try to pace yourself. Questions and problems involving calculations will normally require more time than others and might be weighted more heavily. Wear a watch to keep track of the time and your mental pacing schedule.
8. Answer every question. There is no penalty for guessing.
9. There is an old bromide: If you must guess at the answer, stick with your first hunch. If you erase it and replace it with a second guess, the odds are you will be farther from the scientific truth.

10. Disregard previous answers; that is, if you have guessed (or correctly answered) choice "C" on the previous five questions, do not think that "C" for the next answer is incorrect if you must guess. There is a 20% chance it is correct.

11. **Always** ask yourself once you have selected an answer: *Does My Answer Make Any Sense*? I have seen some absurd answers from people who, when rushed, did not take the time to ask this simple question, yet they could otherwise correctly solve the problem. Wild answers included 154.7 × **10⁶ ppm**, a TWAE of **879 g** of dust/M³, and a **30-min** exposure of a worker to **2300 ppm HCN**, followed by a 7-1/2-hr exposure to the same gas at 0.01 ppm. But then he/she no longer will be working after an inhalation or two. Where is it written that one cannot evaluate the exposure of a corpse?

12. **Memorize** the equations, constants, formulae, atomic weights, conversions, etc. given on the following introductory pages. The equations and constants in boxes are especially important.

13. **Watch your decimal points** and orders of magnitude. **Watch units.** Ensure that they are consistent — 30 thimbles do not a gallon make. Be able to perform a **dimensional analysis** to convert to other units, e.g., from 100 fpm to mph, from 173 mg/sec to tons/day, and 2.3 mg/M³ to lb/ft³:

$$\frac{100\ \text{feet}}{\text{minute}} \times \frac{60\ \text{minutes}}{\text{hour}} \times \frac{\text{mile}}{5280\ \text{feet}} = \frac{1.14\ \text{miles}}{\text{hour}}, \text{ barely a light breeze}$$

$$\frac{173\ \text{mg}}{\text{sec}} \times \frac{60\ \text{sec}}{\text{min}} \times \frac{60\ \text{min}}{\text{hour}} \times \frac{24\ \text{hr}}{\text{day}} \times \frac{\text{g}}{1000\ \text{mg}} \times \frac{\text{lb}}{453.59\ \text{g}} \times \frac{\text{ton}}{2000\ \text{lb}} = \frac{0.0165\ \text{ton}}{\text{day}}$$

$$\frac{2.3\ \text{mg}}{\text{M}^3} \times \frac{\text{M}^3}{35.315\ \text{ft}^3} \times \frac{\text{g}}{1000\ \text{mg}} \times \frac{0.00220\ \text{lb}}{\text{g}} = \frac{1.43 \times 10^{-7}\ \text{lb}}{\text{ft}^3}$$

Please note how fractions are arranged so that identical units cancel each other. A final practice: convert 1.8 mcg/M³/second into lb/ft³/year (= 0.00354 lb/ft³/year).

14. Finally, **Prepare, Don't Pray.** If you want to successfully solve the problems, **Practice, Practice, Practice.** If you work with other industrial hygienists, ask to solve **their** problems. Ask them to share examples of problems which they might have in their notes and professional repertoire.

Dr. Steven Levine of the University of Michigan's School of Public Health offered some of the following tips to prepare for the board certification examinations. Heed his sage advice. Dr. Levine, by the way, is board-certified in both the Comprehensive Aspects and the Chemical Aspects of Industrial Hygiene. The author has augmented Dr. Levine's tips.

AT LEAST SIX MONTHS BEFORE THE EXAMINATIONS

1. Take a comprehensive review course. Refrain from sight-seeing and going "out on the town" while attending this course; instead, study every night. Review the notes for the next day's lectures. Prepare questions for instructors in areas where your concepts and skills are fuzzy.
2. Outline the notes in the book from the comprehensive review course.
3. Condense the outline on 3- × 5-in flip cards. This outlining and condensing "process" will be a valuable learning tool.
4. Buy and use a computerized study and simulated examination program. These programs will give you practice in answering multiple choice questions.
5. Practice every type of calculation you can find. Ten of the most important types of calculations in the area of air contaminants, risk assessment, and ventilation are:

   - "Dr. Clum Z. Chemist" who spilled a bottle of a volatile solvent with known vapor pressure in a room of a given volume with described ventilation parameters
   - Exponential relationships (e.g., radioactive decay, half-value thickness shielding, dilution ventilation, and half-life concentrations)
   - The inverse square law
   - Converting air contaminant concentrations (e.g., converting mg/M^3 into ppm)
   - Vapor pressure calculations
   - Ventilation air volume and hood capture velocity and duct velocity calculations
   - The additive mixture rule for multiple airborne toxicants
   - Time-weighted average dose calculations including consideration of overtime
   - TLVs and PELs for air contaminant mixtures
   - Saturation concentrations in confined, unventilated spaces

   If you understand these basic types of problems, you will be able to do a significant number, perhaps all, of the air sampling types of examination calculations.
6. Exercise during your study breaks. Being physically fit will help you mentally and emotionally and will give you the stamina needed for the one to two days of taking rigorous examinations and tests.
7. Do not reward yourself for wasting study time. For example, if you have a full day to study, and find yourself unable to focus, do "nothing" until you can focus. If, instead, you do other productive work, you will feel good about your alternative productivity, but you will not have accomplished any studying. Every now and then that is O.K.
8. Keep an honest record of those things that you do, and do not, know.
9. Know how to operate your calculators quickly and accurately.

10. Bring an extra calculator (or extra batteries).
11. Study diligently daily. For example, if only three problems are reviewed from this handbook every day, 150 days (@ five months) are needed to review all of them. Handled this way, the study tasks will not be so daunting.

IMMEDIATELY BEFORE AND DURING THE EXAMINATIONS

1. Stay in a comfortable, nearby quiet hotel. Indulge yourself. Consider ordering only nutritionally-balanced room service meals. But do not "hole up" in your room. Get outside once in awhile. Take a brisk walk once or twice a day while making your final preparations.
2. Do not stay with relatives.
3. Do not bring your family, especially any young children.
4. Arrive at the examination location at least one full day early.
5. Eat safe food before the examinations. Consuming alcoholic beverages the night before the examinations is foolish; all of your central nervous system neurons must function at warp speed.
6. Wear comfortably soft, clean, loose-fitting clothing to the examinations. Bring a light sweater in the event the examination room becomes chilly. The author wore a baseball cap when he sat for the examinations. Not only did this help to keep his head warm, but glare was reduced.
7. Arrive before registration starts, get a good seat, relax, take care of any last minute personal needs, and then go back outside of the room to register.
8. Bring bags of candy or other snacks to the examinations so that you can keep your blood sugar constant throughout the day.
9. Bring aspirin, Tylenol®, Motrin®, Excedrin® (which has the caffeine in it), Digel®, and Pepto-Bismol® in a small container to the examinations. While this is not meant to endorse these products, this list covers the full range of over-the-counter drugs to control common, simple problems which might reduce your ability to perform in an optimal fashion.
10. When you get to the examination room, fill out all required forms, and then await the start of the examinations.
11. Spend the next few minutes practicing "positive visualization" where you can see yourself answering the questions accurately, finishing the examinations on time, and then receiving your CIH, framing your certificate, and boasting to others.
12. Do not waste time pondering a single question, thus leaving yourself insufficient time to complete the examinations. Be mindful of your calculated average time allotted for each question.
13. If you finish the examinations early, carefully check every answer (especially your calculations for **decimal points, units, reasonableness of answers**, etc.).

14. Pack an umbrella and/or a raincoat. Sitting for examinations in soggy clothing and with cold, wet hair and shoes might cost a few percentage points off your final score.
15. Clean your glasses. Consider wearing comfortable ear plugs to reduce any noisy distractions. Pack a handkerchief or Kleenex®.

EQUATIONS, CONSTANTS, CONVERSIONS, FORMULAE TO MEMORIZE

Know the empirical formulae of the common alcohols, aromatic hydrocarbons, alkanes, alkenes, ketones, phenols, ethers, etc. and their substituted products (e.g., trichloroethylene, 2-nitropropane, pentachlorophenol, bromobenzene, etc.).

Know the generic, alternative, and trivial names for the common solvents (e.g., methyl chloroform = 1,1,1-trichloroethane, 2-ethoxyethanol = ethylene glycol monoethyl ether, carbon "tet" = carbon tetrachloride, "tri" = trichloroethylene, "perc" = ?, MIBK = ?).

APPROXIMATE ATOMIC WEIGHTS

Memorize
H = 1
C = 12
N = 14
O = 16
S = 32
Cl = 35.5

Na = 23	Cr = 52	Zn = 65	Sb = 122
Si = 32	Fe = 56	As = 75	Hg = 201
Ca = 40	Cu = 64	Cd = 112	Pb = 207

From the above list of approximate atomic weights, be able to calculate molecular weights of common gases, metallic salts, oxides, solvents, etc. For example, without consulting chemistry textbooks and references, calculate the molecular weights of sulfur dioxide, hydrogen cyanide, nitrogen dioxide, acetone, lead carbonate, toluene, chlorine, sulfuric acid, ozone, MEK, zinc oxide, limestone, benzene, carbon monoxide, caustic soda, arsenic trioxide, potash, and trichlorophenol.

GAS AND VAPOR AIR CONTAMINANT MOLECULAR WEIGHTS

Acetic acid	60.05	Isoamyl acetate	130.18
Acetone	58.08	Isoamyl alcohol (n, s)	88.15
Ammonia	17.03	Isobutyl acetate	116.16
Benzene	78.11	Isobutyl alcohol	74.12
2-Butoxyethanol	118.17	Isopropyl alcohol	60.09
Butyl acetate (n, s)	116.16	MDI	250.25
Butyl alcohol (n, s, t)	74.12	Methanol	32.04
Chlorine	70.91	2-Methoxyethanol	76.09
Cyclohexane	84.16	2-Methoxyethanol acetate	118.13
Cyclohexanol	100.16	Methyl chloroform	133.42
Cyclohexanone	98.14	Methylene chloride	84.94
Diacetone alcohol	116.16	Methyl ethyl ketone	72.10
Diisobutyl ketone	142.24	Methyl isobutyl ketone	100.16
Dimethylformamide	73.09	Mineral spirits	≅ 136
Dioxane	88.10	Naphtha (VM and P)	≅ 112
2-Ethoxyethanol	90.12	Nitric oxide	30.01
2-Ethoxyethyl acetate	132.16	Nitrogen dioxide	46.01
Ethyl acetate	88.10	Ozone	48.00
Ethyl alcohol	46.07	Propyl acetate (n)	102.13
Ethyl benzene	106.16	Propyl alcohol (n)	60.09
Formaldehyde	30.03	Styrene	104.14
Gasoline (i.e., ≅ 1–3% **benzene.**)	≅ 73	Sulfur dioxide	64.07
Hexane (n)	86.17	TDI	174.16
2-Hexanone (M n BK)	100.16	Toluene	92.13
Hydrogen bromide	80.92	Trichloroethylene	131.40
Hydrogen chloride	36.46	Triethylamine	101.19
Hydrogen cyanide	27.03	Vinyl acetate	86.09
Hydrogen fluoride	20.01	Vinyl chloride	62.50
Hydrogen sulfide	34.08	Vinyl toluene	118.18
Indene	116.15	Xylene isomers (o, m, p)	106.16

Pandammonium trichloride = ?
2, 5-dimethyl chickenwire = ?
4, 6', 7- β – triphenylawfulstuff = ?
Methyl ethyl death = ?

THE NOT SO MYSTERIOUS MOLE

Moles or mol (the chemical, not the underground kind, nor a FBI or Central Intelligence Agency counterspy) confuse some people. Consider three balloons each containing one mol of a different gas and all at the same temperature and pressure. How can they be so equivalent in most ways, but yet so different? As we see in the table below, other than their molecular chemical properties and mass, each mole is the same as the others. If you know one mol, in a physicochemical sense, you know them all. Moles are not evil animals, but if we "Know thy Beast," we will be better prepared to deal with these chemically-helpful critters.

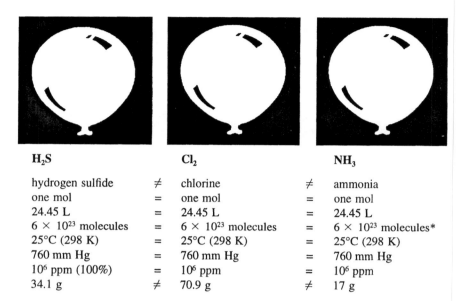

H₂S		**Cl₂**		**NH₃**
hydrogen sulfide	≠	chlorine	≠	ammonia
one mol	=	one mol	=	one mol
24.45 L	=	24.45 L	=	24.45 L
6×10^{23} molecules	=	6×10^{23} molecules	=	6×10^{23} molecules*
25°C (298 K)	=	25°C (298 K)	=	25°C (298 K)
760 mm Hg	=	760 mm Hg	=	760 mm Hg
10^6 ppm (100%)	=	10^6 ppm	=	10^6 ppm
34.1 g	≠	70.9 g	≠	17 g

* Precisely, one mole (one mol) = 6.022045×10^{23} molecules (Avogadro's number — **not** his telephone number which, if it was, and you dialed 100 numbers every second around the clock, it would take 190,827,090,000,000 years to make the call. After three billion years, only 0.0016% of the numbers are dialed, the sun has become a "red giant," and the telephone melts as Earth burns to a cinder.)

Using 24.45 L of H₂S at 760 mm Hg and 25°C as an example, if 34.1 g = 10^6 ppm (100%), then 34.1 micrograms = 1 ppm. Or, we might say:

$$1 \text{ ppm } H_2S = \frac{34.1 \text{ mcg}}{24.45 \text{ L}} = \frac{1.39 \text{ mcg } H_2S}{\text{liter}}$$

Expressed another way: one microliter of H₂S gas diluted to 1 L with pure air = 1 ppm (1 part H₂S gas mixed with 999,999 parts air).

A gram-molecular volume of any gas or vapor occupies 22.4 L at standard conditions of 0°C and 760 mm Hg pressure. This is the weight of 6.022045×10^{23} molecules of that gas or vapor. This assumes that all gases and vapors behave as if they are "ideal" which, for practical purposes at not too deviant temperatures and gas pressures, is the case. Note that the gram-molecular volume becomes 24.45 L at 25°C and 760 mm Hg; that is, heat expands the gas volume at constant pressure and constant mass.

$$\text{ppm} = \frac{\text{mg}}{M^3} \times \frac{22.4}{\text{mol. wt.}} \times \frac{\text{absolute temperature}}{273.15 \text{ K}} \times \frac{760 \text{ mm Hg}}{\text{pressure mm Hg}}$$

at 25°C and 760 mm Hg (at which one gram-mol of a gas or vapor occupies 24.45 L):

$$ppm = \frac{(mcg/L) \times 24.45}{molecular\ weight} \qquad mg/M^3 = \frac{ppm \times molecular\ weight}{24.45}$$

CONVERSION FACTORS AND CONSTANTS

R = 0.0821 L-atm/K-mol   ppm (volume/volume) = $\frac{x}{10^6}$ 1% = 10,000 ppm

mcg = μg = microgram   1 ft^3 ≅ 28.3 L   μL/L = ppm
ppm = micromoles of gas or vapor/mol of gas or vapor   1 mL = 0.001L

K = °C + 273 K $°C = \frac{5}{9}(°F - 32)$ °F = (1.8 × °C) + 32

1 kg = 2.2 lb   1 lb = 454 g   1 in. = 2.54 cm   1 M^3 ≅ 35.3 ft^3
1 L ≅ 0.0353 ft^3 1 in. = 25,400 microns   1 micron = 1 μM = 10^{-6} M
1 gallon = 3.785 L   1 fluid ounce ≅ 29.6 mL   1 oz = 28.35 g
1 gallon H$_2$O ≅ 8.3 lb (at 20°C)   1 gal/min = 0.134 ft^3/min   lb/hr = 10.9 kg/day
mg/L (liquids) = ppm   1 lb/day = 315 mg/min   1 M^3 = 10^6 cM^3 = 10^6 mL
volume of liquid × density (specific gravity) = mass of liquid   area of circle = πr^2
1 atmosphere = 760 mm Hg = 29.9-in Hg = 14.7 lb/in^2 volume of cylinder = πr^{2}h
volume of sphere = 1.333πr^3 1 M^3 = 1000 L   20 "drops" ≅ 1 mL

IDEAL GAS (AND VAPOR) LAWS

GENERAL GAS LAW: **P V = n R T**

n = number of mols   R = universal gas constant = 0.0821 L-atm/mol-K
i = initial gas condition   f = final gas condition
P = gas pressure (atm)   V = gas volume (L)

T = **absolute** temperature (K) — Read: **Absolute** temperature.

$$\frac{P_i V_i}{T_i} = \frac{P_f V_f}{T_f}$$

rearranging:

$$V_f = \frac{V_i T_f P_i}{T_i P_f}$$

Charles' law: The volume of a mass of gas is directly proportional to its temperature on the Kelvin scale when the pressure is held constant, or:

$$\frac{V_i}{T_i} = \frac{V_f}{T_f}$$

Boyle's law: The volume of a mass of gas held at constant temperature is inversely proportional to the pressure under which it is measured, or:

$$\frac{P_i}{V_i} = \frac{P_f}{V_f}$$

Daltons' law of partial pressures: The total pressure of a mixture of gases is equal to the sum of partial pressures of the component gases and vapors.

Gay-Lussac's law: At constant volume, the pressure of a mass of gas varies directly with the absolute temperature, or:

$$\frac{P_i}{T_i} = \frac{P_f}{T_f}$$

Raoult's law: The partial vapor pressure of each component of a solvent mixture equals its vapor pressure multiplied by its mole fraction in the liquid mixture, i.e., a solution's vapor pressure is proportional to the mole fraction of its component solvents.

$$V_o = \frac{298 \times V \times P}{760 \times T}$$

where V_o = standard air volume in liters, V = the indigenous volume of sampled air in liters, P = ambient pressure in mm of Hg (note: pressure is the actual barometric pressure adjusted to sea level), and T = the ambient temperature in kelvin, $K = 273$ K + ambient temperature, °C.

In order to compare data on gases and vapors for direct-reading instruments with air quality standards, the meter reading in ppm must be converted to ppm at a normal temperature and pressure (NTP = 25°C and 760 mm Hg) by using the formula:

$$\text{ppm (at NTP)} = \text{ppm}_{meter} \times \frac{P}{760 \text{ mm Hg}} \times \frac{298 \text{ K}}{T}$$

where:

ppm_{meter} = meter reading in ppm
P = sampling site barometric pressure (in mm Hg)
T = sampling site air temperature (in kelvin, K)

Barometric pressure is obtained by checking a calibrated barometer or calling the local weather station or the airport. Ask for the unadjusted barometric pressure. If these sources are unavailable, then a good "rule of thumb" is: for every 1000 ft of elevation, the barometric pressure decreases by approximately one in of Hg.

DILUTION VENTILATION AND TANK AND ROOM PURGING

Industrial hygienists are frequently called upon to predict the concentration of an air contaminant remaining in a room or space after operation of an exhaust or air supply ventilation system. Dilution of air contaminants follows first order exponential decay kinetics if there is good mixing of dilution air with contaminated air. The key equation for these calculations is:

$$C = C_o e^{-\left[\frac{Q}{V}\right]t}$$

where:

C = concentration of contaminant remaining after operating the ventilation system for a specified time period, t, usually in minutes

C_o = the original air contaminant concentration, usually in ppm or mg/M^3

Q = ventilation rate [supply or exhaust, but not both (take the larger of the two) — usually in cubic feet of air/min, cfm]

V = volume of the room or space, usually in cubic feet

k, a ventilation mixing factor (with 1 = perfect and 10 = extremely poor), is a subjective rating based on experience and judgment and applied as a safety factor.

VENTILATION PURGE TIME EQUATION

This equation is used to calculate the time required to reduce the concentration of an air contaminant using ventilation in a space of known volume. Depending on the air mixing characteristics, application of safety factors requiring professional judgment is required.

$$t = \frac{-\ln\left[\dfrac{C}{C_o}\right]}{\dfrac{Q}{V}}$$

where:

C = concentration after time t (minutes)

C_o = original concentration

Q = cfm

V = volume (ft^3) of the space (room, tank, railroad boxcar, industrial plant, etc.).

This equation assumes perfect mixing of the dilution air with the contaminated air (k = 1.0). Otherwise, k = >1 to 10. In other words, if mixing is less than perfect (k, say = 5), and 30 min are required for dilution if mixing is perfect, 150 min are required in this example. Know the 10% rule and the 50% dilution ventilation rules:

2.3, 4.6, and 6.9 chamber, or room, volumes of clean air are needed, respectively, to dilute an air contaminant to 10%, 1%, and 0.1% of its initial concentration. This assumes the perfect mixing of clean air with the contaminated air and no further generation of the air contaminant as the ventilation proceeds.

Under the same conditions, after a volume of air equal to 50% of the room or chamber volume has mixed with contaminated air, the original concentration will have been reduced by nearly 40% (to 60.7% of the original concentration).

EXAMPLE

If C_o = 1000 ppm, Q = 230 cfm, and V = 1000 ft³, how long will it take to dilute the vapor or gas concentration to 100 ppm? Assume perfect mixing.

$$t = \frac{-\ln\left[\dfrac{100 \text{ ppm}}{1000 \text{ ppm}}\right]}{\dfrac{230 \text{ ft}^3/\text{minute}}{1000 \text{ ft}^3}} = \frac{-\ln 0.1}{0.23} = \frac{-(-2.3)}{0.23} = 10 \text{ minutes}$$

$$\text{time to dilute to 1 ppm} = t = \frac{-\ln\left[\dfrac{1 \text{ ppm}}{1000 \text{ ppm}}\right]}{\dfrac{230 \text{ cfm}}{1000 \text{ ft}^3}} = \frac{-(-6.9)}{0.23} = 30 \text{ minutes}$$

EXAMPLE

C_o = 1000 ppm
V = 1000 ft³ (e.g., 10 × 10 × 10 ft)
Q = 230 cfm
t = 10, 20, 30, and 40 min (or 2.3, 4.6, 6.9, and 9.2 room volumes of clean dilution air)

after 10 min:

$$C = C_o e^{-\left[\frac{Q}{V}\right]t} = (1000 \text{ ppm}) e^{-\left[\frac{230 \text{ cfm}}{1000 \text{ ft}^3}\right]10 \text{ minutes}} = (1000 \text{ ppm}) e^{-2.3} = 100 \text{ ppm}$$

after 20 min:

$$C = (1000 \text{ ppm}) e^{-\left[\frac{230 \text{ cfm}}{1000 \text{ ft}^3}\right]20 \text{ minutes}} = (1000 \text{ ppm}) e^{-4.6} = 10 \text{ ppm}$$

after 30 min: C = 1 ppm
after 40 min: C = 0.1 ppm
Note that the concentrations are reduced to 10% of the original concentrations for every **2.3** chamber dilution volumes.

$$C = C_o e^{\frac{q}{V}}$$

where:

C = resultant concentration
C_o = original concentration
q = withdrawn sample volume
V = chamber or room volume.

EXAMPLE using a 50% ventilation volume:

C_o= 1000 ppm
V = 1000 ft^3

q = 500 ft^3

$$C = (1000 \text{ ppm}) \, e^{-\frac{500 \text{ ft}^3}{1000 \text{ ft}^3}} = (1000 \text{ ppm}) \, e^{-0.5} = (1000 \text{ ppm}) (0.6065) = 607 \text{ ppm}$$

Using $\log_{10}$:

$$\frac{q}{V} = 2.3 \times \log\left[\frac{C_o}{C}\right]$$

Use

$$\frac{C}{C_o} = e^{-\left[\frac{Q}{V}\right]t}$$

to determine the fraction remaining after operating the ventilation system for a specified time.

AIR CONTAMINANT HALF-LIFE

An air contaminant half-life, $C_{1/2}$ (the time required to dilute an air contaminant to 50% of its original concentration), is calculated by:

$$C_{1/2} = 0.693\left[\frac{V}{Q}\right]$$

where:

V = volume of room, space, or container and
Q = the uniform dilution ventilation rate.

For example, a 1000 ppm gas or vapor concentration is reduced to 500 ppm after about 14 min by 200 cfm in a room $20 \times 20 \times 10$ ft (4000 ft³), to 250 ppm after 28 min, and to 125 ppm after 42 min, i.e., $0.693 \left(\dfrac{4000 \text{ ft}^3}{200 \text{ ft}^3/\text{minute}} \right) = 13.86$ minutes. This applies when there is no further generation of the air contaminant, and there is excellent mixing of uncontaminated air with the polluted air.

DENSITY (SPECIFIC GRAVITY) CALCULATIONS

The term density denotes the **ratio** of the mass of a substance to that of an equal volume of a reference substance (usually water at a specified temperature). Density is essentially identical to the specific gravity of a substance, the terms being normally interchangeable. For gases and vapors, air is often used as the reference.

Water at 4°C is at its maximum density = 1.000 g/cm³ = 1.000 g/mL = 62.4 lb/ft³ = 1 kg/L (= 0.997 g/cm³ at 25°C). Water is anomalous among liquids since it is denser at 4°C than at its freezing point of 0°C (Otherwise ice would sink to the bottom of the lake.).

Density calculations are used to determine the mass of a liquid volume. For example, the industrial hygienist might need to calculate the mass of a contaminant that evaporates from the spill of a specified volume of volatile liquid. To do this:

liquid volume $\times$ liquid's density = mass of liquid

For example, 50 mL of toluene totally evaporate after being spilled upon the floor:

$$50 \text{ mL of liquid phase toluene} \times \frac{0.87 \text{ g}}{\text{mL}} = 43.5 \text{ grams of vapor phase toluene}$$

Hydrocarbons which are not substituted with halogens (e.g., chlorine, bromine, fluorine) generally have densities less than water (i.e., < 1g/mL). That is, they float on water (e.g., oil, paint thinner, gasoline) if they do not appreciably mix or dissolve in the water. Example densities (in g/mL, at normal temperature) are:

kerosene	0.82	benzene	0.88	ethanol	0.79
acetone	0.79	methanol	0.79	toluene	0.87
n-butanol	0.81	CH_2Cl_2	1.34	CCl_4	1.59

Acetone, methanol, ethanol, and butanol will mix with water. Kerosene, toluene, and benzene do not and, since their densities are less than water, they float on water. Carbon tetrachloride and methylene chloride, with densities greater than water, will sink below water since they do not appreciably dissolve in water.

VAPOR PRESSURE PROBLEMS AND CALCULATIONS

When the vapor pressure of a liquid or volatile solid is known, one can calculate the maximum, or saturation, vapor concentration which can exist in an enclosed space. Such calculations are very helpful when the industrial hygienist wishes to determine the worst possible scenario, and air sampling instruments are not readily available. In the case of a closed vessel containing a volatile material, such as a chemical storage tank, molecules will escape (i.e., evaporate, volatilize) into air or gas space above the liquid until eventually the enclosed atmosphere above the liquid can no longer hold any more vapor molecules at that pressure and temperature. The air (or other gas) is saturated at this point with molecules of the evaporating liquid (or certain solids which sublime).

Keep in mind that many solids have significant vapor pressures as well. Iodine crystals, naphthalene flakes, p-dichlorobenzene, organophosphate pesticides, DDT, and phthalates are examples of solids which, sooner or later, will evaporate. It should be obvious that liquids, in general, have higher vapor pressures than solids and that there is a tremendous range of vapor pressures between different chemicals. Again, many solids exhibit substantial vapor pressures at room temperature (e.g., iodine, dichlorobenzene, naphthalene, dichlorobenzene, phthalates, phenols).

One should not be misled that a chemical with a lower vapor pressure, all other parameters being equal, is not necessarily less hazardous than another with a higher vapor pressure. A chemical with a low vapor pressure, when spread over a large area, can present a greater inhalation exposure hazard than a highly volatile material with a small surface area (See problem 375).

To calculate the maximum vapor concentration of a volatile material (at usually an assumed temperature of 25°C and 760 mm Hg):

$$\frac{\text{vapor pressure at a temperature}}{\text{barometric pressure}} \times 10^6 = \text{ppm of vapor}$$

For example, the vapor pressure of 2-nitropropane is 13 mm Hg at 20°C. What is the saturation concentration of vapor in a tank containing liquid 2-NP where the barometric pressure is 710 mm Hg?

$$\frac{13 \text{ mm Hg}}{710 \text{ mm Hg}} \times 10^6 = 17,105 \text{ ppm } 2-\text{NP}$$

This concentration of 2-nitropropane vapor (1.71%) is considerably greater than the ACGIH TLV ceiling concentration of 10 ppm (A 3 carcinogen), but it is slightly below the LEL of 2.6%. Inhalation of this vapor concentration would be a significant health hazard and, since the concentration is greater than 10–20% of the LEL, this also presents a potential fire and explosion hazard. The NIOSH IDLH for 2-NP is 100 ppm.

Some trivia: 1 ppm ≈ 1 in. in 16 miles

1 ppb ≈ 1 in. in 16,000 miles ≈ 4 in² in a square mile

Dry air in a standard sewing thimble at sea level weighs about 3600 micrograms. A standard office staple ≅ 35,000 micrograms. Contrast this with the OSHA PEL of 50 mcg/M³ for lead dust, fume, and mist aerosols (See problem 245).

ABIH CERTIFICATION EXAMINATION HOMEWORK: AIR CALCULATIONS AND VENTILATION

The following 14 problems are considered to be the most basic and fundamental to an understanding and comprehension of principles of industrial hygiene air sampling and ventilation. To prepare for the American Board of Industrial Hygiene certification examinations in the comprehensive practice of industrial hygiene, solve the following problems. If one understands and can solve these problems, they are in an excellent position to handle many of the related problems presented in the ABIH examinations.

The rules are:

1. Do not look at the following problems until you are ready to solve them. Work alone. Do not solicit help.
2. Do not use books, notes, or other reference sources.
3. Use only pencils, a nonprogrammable calculator, a wrist watch, and a note pad.
4. Solve as many problems as you can in one hr. (This is an average of 4.3 min per problem.)
5. Compare your answers to the correct answers in the book as given by the problem numbers in parentheses.

Good Luck!

1. BZ air was sampled for total barley dust at 1.8 L/m for five hrs, 40 min with a 37 mm MCE MF with respective pre-sampling and post-sampling weights of 33.19 mg and 38.94 mg. What was the grain silo filler's eight-hr TWAE exposure to respirable dust if 85 mass-percent was nonrespirable? (Problem 304)
2. Air was sampled for HCl gas (mw = 36.45) in 15 mL of impinger solution at 0.84 L/m for 17 min, 20 sec. HCl collection efficiency was 80%. A chemist analyzed 4.7 mcg Cl/mL in the sample and 0.3 mcg/mL in the blank impinger. What was the steel pickler's exposure in ppm? (Problem 305)
3. Determine eight-hr TWAE of a scrap metal processor to Pb dust and fume with exposures of three hrs, 15 min to 17 mcg Pb/M³; 97 min to 565 mcg Pb/M³; and two hrs, 10 min to 46 mcg Pb/M³. The worker wore an approved HEPA dust/fume/mist filter cartridge respirator for 5 1/2 hrs. (Problem 306)
4. 172 g of liquid phosgene splash onto a floor. What gas volume quickly results after evaporation at an air temperature of 23°C and an atmospheric pressure of 742 mm Hg? COCl₂ molecular weight = 99. Boiling point of phosgene = 47°F. (Problem 140)

5. A chemical plant operator had the following eight-hr TWAEs on Monday: 32 ppm toluene, 19 ppm xylene, and 148 ppm MEK. Their respective TLVs are 50, 100, and 200 ppm. By what percent is the additive exposure limit exceeded? (Problem 308)

6. Air in an empty room (20 × 38 × 12 ft) contains 600 ppm cyclohexene vapor. How long will it take to dilute this to 6 ppm with a 1550 cfm vane-axial exhaust fan? K factor = 3. (Problem 309)

7. 7.3 µL liquid styrene (mw = 104.2, density = 0.91 g/mL) are evaporated in a 21.6 L glass calibration bottle. What is the styrene vapor concentration in ppm? (Problem 310)

8. A rotameter was calibrated at 25°C and 760 mm Hg. What is the corrected air flow rate when the rotameter indicates 2.0 L/m at 630 mm Hg and 33°C? (Problem 311)

9. What is the effective specific gravity of 13,000 ppm of a gas in air when the gas has a specific gravity of 4.6? Will the mixture stratify with the denser gas at floor level? (Problem 312)

10. Analysis of an 866 L MF air sample detected 2667 mcg Zn. How much zinc oxide (ZnO) fume does this represent in the welder's breathing zone? (mw Zn and O = 65 and 16, respectively) (Problem 313)

11. What is the air flow rate through an 8-in diameter unflanged duct with a transport velocity of 2900 fpm? What capture velocity is expected 8 in. in front of the duct inlet? Without cross-drafts, what discharge velocity can be expected 20 ft from the exhaust outlet? What if the exhaust duct inlet has a wide flange? (Problem 314)

12. A 47 × 166 × 20-ft building is supplied with 7300 cfm. How many air changes occur per hr? How many minutes are required per air change? How many cubic feet are supplied per square foot of floor area (20 ft is the height of the ceiling)? Forty-seven people work in this single-story building. What is the outdoor air ventilation rate per person if 90% of the air is recirculated? (Problem 271)

13. An exhaust system operates at 19,400 cfm. A hood is added to the system which requires a total system capacity of 23,700 cfm. By how much should fan speed be increased to handle the extra exhaust volume? (Problem 276)

14. In the preceding problem, what is the required increase in fan horsepower to handle the increased air volume? (Problem 277)

Handbook Problems

1. Air was sampled through a 37 mm PVC membrane filter for seven hrs and 37 min. The initial air flow rate was checked twice with a 1000 mL soap film bubble calibrator at 49.7 and 50.1 sec per liter. Post-calibrations were 53.7 and 53.3 sec per liter. What volume of air (at standard conditions of 25°C and 760 mm Hg) was sampled in liters and in cubic feet?

7 hrs and 37 min = 420 min + 37 min = 457 min

$$\frac{(49.7+50.1+53.7+53.3) \text{ seconds}}{4} = 51.7 \text{ seconds}$$

$$\text{liters/minute} = \frac{60 \text{ seconds/liter}}{51.7 \text{ seconds/liter}} = 1.16 \text{ L/min} \quad 530 \text{ L} \times \frac{1 \text{ ft}^3}{28.3 \text{ L}} = 18.7 \text{ ft}^3$$

Answers: 530 L. 18.7 ft³.

2. A solvent degreaser operator's exposures to trichloroethylene vapor were 2 1/4 hrs at 57 ppm, 3 1/2 hrs at 12 ppm, 1 3/4 hrs at 126 ppm, and 30 min at 261 ppm on one work day. What was her TWAE? Did it exceed the TLV? The AL?

Haber's law = concentration × time = CT   8-hr TWAE worker dose calculation

$$
\begin{array}{llll}
57 \text{ ppm} & \times & 2.25 \text{ hrs} = & 128 \text{ ppm-hrs} \\
12 \text{ ppm} & \times & 3.5 \text{ hrs} = & 42 \text{ ppm-hrs} \\
126 \text{ ppm} & \times & 1.75 \text{ hrs} = & 221 \text{ ppm-hrs} \\
261 \text{ ppm} & \times & 0.5 \text{ hr} = & 131 \text{ ppm-hrs} \\
& & 8.0 \text{ hrs} = & 522 \text{ ppm-hrs}
\end{array}
$$

$$\frac{522 \text{ ppm-hours}}{8.0 \text{ hours}} = 65 \text{ ppm TWAE to TCE vapor}$$

Answers: TWAE = 65 ppm. Yes, by 15 ppm. Yes, 2.6 × the AL of 25 ppm.

3. In the previous problem, assume that the worker wore an organic vapor cartridge respirator with an overall efficiency of 90% (filtration efficiency + face-to-mask seal). What was the face piece penetration? What was the respirator protection factor? What was her true TWAE assuming equal protection at all vapor concentrations?

$$\text{respirator protection factor} = \frac{\text{ambient concentration}}{\text{concentration inside facepiece}}$$

65 ppm $\times$ 0.9 = 58.5 ppm (i.e., 90% of 65 ppm)
65 ppm – 58.5 ppm = 6.5 ppm penetrated the respirator (10% penetration)

$$\frac{65 \text{ ppm}}{6.5 \text{ ppm}} = PF = 10$$

Answers: 10%. Protection factor = 10. TWAE $\cong$ 7 ppm.

4. What volume will 73 g of dry ammonia gas occupy at 11°C and 720 mm Hg? How much air is needed to dilute the ammonia to 10 ppm?

$$PV = nRT \quad V = \frac{nRT}{P} \quad R = 0.0821 \text{ L-atm/mole-K}$$

molecular weight $NH_3 = 17 \quad n = \dfrac{73 \text{ g}}{17 \text{ g/mole}} = 4.29 \text{ moles}$

T = 273 K + 11°C = 284 K   P = 720 mm Hg/760 mm Hg = 0.947 atmosphere

$$V = \frac{(4.29 \text{ moles } NH_3)(0.0821)(284 \text{ K})}{0.947 \text{ atmosphere}} = 105.5 L \quad 10 \text{ ppm} = \frac{10}{10^6} = \frac{1}{10^5}$$

Answers: 105.5 L. 105.5 $\times$ 10^5 L of air which dilutes 100% to 0.001%.

5. What is the cumulative error of several measurements if the day of air sampling was $\pm$ 50% of the true daily exposure, analytical accuracy was $\pm$ 10% of the true value, air sample timing was $\pm$ 1% of the true value, and the air flow rate error was $\pm$ 5% of the true value?

cumulative error, $E_c = \pm\sqrt{(E_1)^2 + (E_2)^2 + \ldots (E_n)^2}$

$$E_c = \pm\sqrt{50^2 + 10^2 + 1^2 + 5^2} = \pm\sqrt{2626} = \pm51.2\%$$

Answer: $\pm$ 51.2% of the true value. This is referred to as the sampling analytical error (SAE) which is used to calculate the lower and upper confidence limits for air sampling test results (See problem 430).

6. A paint sprayer had TWAEs to MEK of 68 ppm, toluene of 37 ppm, n-butyl alcohol of 6 ppm, and xylene of 23 ppm, all as vapor phase air contaminants and not as mist. Assume effects are toxicologically additive and that he did not wear a respirator. What was his equivalent exposure? OSHA PELs are 200, 100, 50 (C), and 100 ppm, respectively.

in ppm: $\dfrac{\text{exposure}}{\text{respective PEL or TLV}} + \ldots \dfrac{E_n}{PEL_n}$, in TWAEs

$$\frac{68}{200}+\frac{37}{100}+\frac{6}{50}+\frac{23}{100}=1.06\,(\text{no units})$$

Answer: 1.06, or 6% above PEL for the mixture and 2.1 times the Action Level of 0.5.

7. A polyurethane foam machine operator had TWAEs to vapors of TDI of 0.003 ppm and CH_2Cl_2 of 36 ppm. What was his equivalent exposure to the air contaminant mixture?

   Answer: Generally, the additive mixture rule would not apply. Although both are irritants to the respiratory tract and mucous membranes, TDI is a sensitizer, and CH_2Cl_2 primarily affects the CNS and blood HgB (COHb formation) and potentially is a liver and lung carcinogen. TDI is a potential carcinogen.

8. What is the specific gravity of a mixture of 10,000 ppm Cl_2 gas in dry air? The specific gravity of Cl_2 gas is 2.5.

   10,000 ppm (volume/volume) = 1%

   for air: $0.99 \times 1.0 = 0.990$
   for Cl_2: $\underline{0.01} \times 2.5 = \underline{0.025}$
   $1.00 1.015$ (air = 1.000)

   Answer: Specific gravity (or relative density) = 1.015, or only 1.5% greater than air. In a practical sense with respect to designing work place ventilation, there is no great difference between air and a very high concentration of vapors or gases at, approximately, 1–5% (e.g., 2.5 times heavier than air).

9. An industrial hygiene chemist analyzed 0.256 mg Zn on an air filter used for a 96 L air BZ sample. What was the welder's exposure to zinc oxide fume during the air sampling period?

   molecular weights: Zn = 65, O = 16, ZnO = 81

   $$\frac{65}{81}\times100=80\%\text{ Zn in ZnO}\qquad\frac{0.256\text{ mg Zn}}{0.80}=0.32\text{ mg ZnO}=320\text{ mcg ZnO fume}$$

   $$\frac{\text{mcg}}{\text{L}}=\frac{\text{mg}}{\text{M}^3}\qquad\frac{320\text{ mcg ZnO}}{96\text{ L}}=3.3\text{ mg ZnO/M}^3$$

   Answer: 3.3 mg zinc oxide fume/M^3.

10. A one quart bottle of *n*-butanol broke upon falling to the floor entirely evaporating in a $10 \times 40 \times 80$-ft room with 10% room contents. What average vapor concentration exists assuming there is no ventilation? *n*-Butanol density = 0.81 g/mL.

$(10 \times 40 \times 80 \text{ ft}) - 10\% = 32,000 \text{ ft}^3 - 3200 \text{ ft}^3 = 28,800 \text{ ft}^3$

$$\frac{28,800 \text{ ft}^3}{35.3 \text{ ft}^3/\text{M}^3} = 816 \text{ M}^3 = \text{net volume of room}$$

1 quart = 1/4 (3785 mL/gal) = 946 mL $\times$ 0.81 g/mL = 766 g of liquid n-butanol eventually evaporates into the room air

$CH_3(CH_2)_2CH_2OH = C_4H_{10}O = 4 + 810 + 16 = \text{molecular weight} = 74$

$$\text{ppm} = \frac{\dfrac{766,000 \text{ mg}}{816 \text{ M}^3} \times 24.45}{74} = 310 \text{ ppm n-butanol vapor}$$

Answer: 310 ppm n-butanol vapor.

11. What can we conclude from the previous problem?

    a. A fire hazard, but no health hazard
    b. A health hazard, but no fire hazard
    c. A fire hazard and a health hazard
    d. No fire or health hazard
    e. A combustion risk, but not a flammability hazard
    f. None of the above

    Answer: b. n-Butanol TLV and PEL = 50 ppm (C) and Skin notation. LEL – UEL = 1.4 to 11.2%. 310 ppm = 0.031%. Therefore, less than the LEL but greater than the PEL. 620% of PEL and 2.2% of LEL.

12. Salt mine air was sampled through a PVC filter which initially weighed 73.67 mg and 88.43 mg after sampling. The initial air flow rate was 2.18 L/min; the final flow rate was 1.98 L/min after seven hrs and 37 min. The analyzed filter had 5.19 mg sodium after blank correction. What were the concentrations of salt dust and total dust in the air sample?

    7 hrs and 37 min = 420 min + 37 min = 457 min

    88.43 mg – 73.67 mg = 14.76 mg

    $$\text{L/min} = \frac{(2.18 + 1.98)\text{L/min}}{2} = 2.08 \text{ L/min average}$$

    457 min $\times$ 2.08 L/min = 951 L of air sampled

    $$\frac{14.76 \text{ mg}}{951 \text{ L}} = \frac{14.76 \text{ mg}}{0.951 \text{ M}^3} = 15.5 \text{ mg total dust/M}^3$$

    molecular weights: Na = 23; Cl = 35.5; NaCl = 58.5

    58.5/23 = 2.54   5.19 mg Na $\times$ 2.54 = 13.2 mg NaCl

$$\frac{13.2 \text{ mg NaCl}}{0.951 \text{ M}^3} = 13.9 \text{ mg NaCl/M}^3$$

difference = 15.5 mg/M³ – 13.9 mg/M³ = 1.6 mg/M³. Could this be diesel engine exhaust smoke from the underground mining equipment?

Answers: 13.9 mg NaCl/M³. 15.5 milligrams of total dust, smoke, and fume/M³.

13. A closed 100,000 gallon storage tank in Houston contains 10,000 gallons toluene. What is the equilibrium saturation vapor concentration in the tank at 20°C? The vapor pressure of toluene at 20°C = 22 mm Hg.

Since Houston is close to sea level, assume the barometric pressure = 760 mm Hg.

$$\frac{22 \text{ mm Hg}}{760 \text{ mm Hg}} \times 10^6 = 28,947 \text{ ppm} = 2.89\% \text{ (volume/volume)}$$

Answer: ≅ 29,000 ppm (2.9%).

14. In the preceding problem, is the tank atmosphere explosive? The LEL and UEL for toluene = 1.2 and 7.1%, respectively.

Answer: Yes. Watch out. 2.9% exceeds LEL, but less than UEL. This is most hazardous when vapors are in the stoichiometric mid-range of explosibility. This tank "carries its own match." Control ignition sources. Consider the use of inert gas to reduce O_2 concentration (e.g., to < 6% O_2). Ventilate before entry, confined space entry practices, train workers, air testing, post and label, etc.

15. In problem 13, how many pounds of toluene are in the vapor phase?

$$22.4 \text{ L/gram-mole} \times \frac{273 \text{ K} + 20°\text{C}}{273 \text{ K}} = 24.04 \text{ L/gram-mole}$$

toluene = C_7H_8 (7 × 12) + (8 × 1) = molecular weight = 92

$$mg/M^3 = \frac{ppm \times mol. \text{ wt.}}{24.04} = \frac{28,947 \times 92}{24.04} = 110,779 \text{ mg/M}^3 = 110.8 \text{ grams/M}^3$$

100,000 gallon tank – 10,000 gallons liquid C_7H_8 = 90,000 gallons air-vapor space

90,000 gallons × 3.785 L/gallon = 340,650 L = 340.65 M³

340.65 M³ × 110.8 g/M³ = 37,744 g of vapor phase toluene

$$\frac{37,744 \text{ grams}}{454 \text{ grams/lb}} = 83.1 \text{ lb}$$

Answer: There are 83.1 pounds of vapor phase toluene in the tank.

16. In problem 13, what is the toluene vapor concentration after 45 min operation of a 2000 cfm dilution blower? Assume good mixing of fresh dilution air or nitrogen with the toluene-contaminated air and there is negligible evaporation of toluene as the ventilation proceeds.

$$C = C_0 e^{-\left[\frac{Q}{V}\right]t} = \text{ resultant air contaminant concentration}$$

90,000 gallons $\times$ 3.785 L/gal = 340,650 L = 12,037 ft^3 C = (28,947 ppm) $\times$

$$e^{-\left[\frac{2000 \text{ cfm}}{12,037 \text{ ft}^3}\right] \times 45 \text{ minutes}} = (28,947 \text{ ppm}) e^{-7.48} = 28,947 \text{ ppm} \times 0.00056 = 16.3 \text{ ppm}$$

Answer: $\cong$ 16 ppm toluene vapor assuming static conditions and negligible evaporation of toluene during dilution ventilation. The liquid toluene present would evaporate as ventilation commenced. Special dilution ventilation calculations can be used if the vaporization rate is known (see ACGIH's *Industrial Ventilation*). The use of air to initially ventilate could be very hazardous, especially if there is a fan which generates spark or heat of air friction. The better part of valor might be to reduce the oxygen concentration in the head space to less than 6% by volume with a nitrogen gas purge and then ventilate with air to reduce toluene vapor concentration to << TLV/PEL.

17. Assume that the toluene solvent storage tank in problem 13 is located in the Rocky Mountains where the barometric pressure is 640 mm Hg. What is the saturation vapor concentration?

$$\frac{22 \text{ mm Hg}}{640 \text{ mm Hg}} \times 10^6 = 34,375 \text{ ppm} = 3.44\% \text{ (volume/volume)}$$

Answer: 34,375 ppm toluene vapor exceeds the LEL, but is below the UEL. However, the LEL-UEL limits change with changes in the partial pressure of O_2, i.e., altitude effects can alter the LEL-UEL range. Changes in temperature also affect the LEL and UEL range. Increased temperature and oxygen lower the LEL.

18. A solvent is 2% (v/v) benzene and 98% (v/v) toluene. What are the percent vapor phase concentrations of each component? The vapor pressures of benzene and toluene are 75 mm Hg and 22 mm Hg, respectively. The densities of benzene and toluene are 0.88 g/mL and 0.867 g/mL, respectively.

Use 100 mL of the solvent mixture as the volume basis of Raoult's law calculations:

2 mL × 0.88 g/mL = 1.76 g of benzene/100 mL   molecular weight = 78.1

98 mL × 0.867 g/mL = 85 g of toluene/100 mL   molecular weight = 92.1

$$\text{partial vapor pressure from benzene} = \frac{\dfrac{1.76\ g}{78.1\ g/mole} \times 75\ mm\ Hg}{\dfrac{1.76\ g}{78.1\ g/mole} + \dfrac{85\ g}{92.1\ g/mole}} = 1.79\ mm\ Hg$$

$$\text{partial vapor pressure from toluene} = \frac{\dfrac{85\ g}{92.1\ g/mole} \times 22\ mm\ Hg}{\dfrac{1.76\ g}{78.1\ g/mole} + \dfrac{85\ g}{92.1\ g/mole}} = 21.47\ mm\ Hg$$

total saturation vapor pressure = 1.79 mm Hg + 21.47 mm Hg = 23.26 mm Hg

$$\frac{1.79\ mm\ Hg}{23.26\ mm\ Hg} \times 100 = 7.7\%\ \text{benzene vapor}$$

$$\frac{21.47\ mm\ Hg}{23.26\ mm\ Hg} \times 100 = 92.3\%\ \text{toluene vapor}$$

Answers: 7.7% benzene vapor and 92.3% toluene vapor. Calculations were made using Raoult's law which demonstrates enrichment in the vapor phase by the more volatile component, i.e., note how benzene enriched from 2% in the liquid phase to 7.7% (3.9 times) in the vapor phase. The following quotation from Harris and Arp (*Patty's Industrial Hygiene and Toxicology*, Third Edition) is noteworthy:

> Raoult's law should be used with caution in estimating emissions from partial evaporation of mixtures; not all mixtures behave as perfect solutions. Elkins, Comproni, and Pagnotto measured benzene vapor yielded by partial evaporation of mixtures of benzene with various aliphatic hydrocarbons, chlorinated hydrocarbons, and common esters, as well as partial evaporation of naphthas containing benzene. Most measurements for all four types of mixtures showed greater concentrations of benzene in air than were predicted by Raoult's law. Of five tests with naphtha-based rubber cements, one yielded measured values of benzene concentration in air in agreement with calculated values, the other four showed measured benzene concentrations in air to be 3–10 times greater than those calculated using Raoult's law.

> Substantial deviation from Raoult's law is not always the case, however, even with benzene. Runion compared measured and calculated concentrations in air of benzene in vapor mixtures yielded by evaporation from a number of motor gasolines and found excellent agreement.

19. A mine atmosphere averages 12 mg total dust/M³ of air. If this dust is 9% mass respirable and has 8% crystalline quartz in the respirable fraction, how long must one sample at 1.7 L/m if analytical sensitivity is 50 mcg of α-quartz?

    12 mg total dust/M³ = 12 mcg/L (12 mcg/L) × 0.9 × 0.8 = 0.0864 mcg/L

$$\frac{50 \text{ mcg quartz}}{0.0864 \text{ mcg quartz/L}} = 578 \text{ L} \qquad \frac{578 \text{ L}}{1.7 \text{ L/m}} = 340 \text{ minutes}$$

    Answer: 340 min = 5 hrs and 40 min.

20. Two impingers are connected in series. Calculate the collection efficiency of the first impinger if it contains 78.9 mcg, and the second impinger contains 6.3 mcg of the same air contaminant.

$$\% \text{ efficiency} = 100 \left[1 - \frac{C_2}{C_1} \right] = 100 \left[1 - \frac{6.3 \text{ mcg}}{78.9 \text{ mcg}} \right] = 100 \left(1 - 0.08 \right) = 92\%$$

    Answer: 92% of the total air contaminant is in the first impinger assuming both impingers collected 100%. A correction factor of 1.08 could be applied to the concentration found in the first impinger (i.e., 92/(78.9 + 6.3) = 1.08).

21. 6.7 ft³ of air at 53°F and 14.7 lb/in² are adiabatically compressed to 95 lb/in². What is the initial temperature of the air after compression? What is the final volume of the compressed air? 1.4 = the specific heat (the ratio of heat capacity at constant pressure to the heat capacity at constant volume, often expressed as k).

    460 + 53°F = 513°R absolute temperature

$$\text{temperature} = 513° \left[\frac{95}{14.7} \right]^{\frac{1.4-1}{1.4}} = 875°R = 415°F$$

$$\text{volume} = 6.7 \text{ ft}^3 \left[\frac{14.7}{95} \right]^{\frac{1}{1.4}} = 6.7 \text{ ft}^3 (0.155)^{0.714} = 1.77 \text{ ft}^3$$

    Answers: 415°F and 1.77 ft³. The value of k is a function of temperature and pressure. For air and several diatomic gases (e.g., N_2, O_2), k equals 1.4. Many hydrocarbons have k values typically between 1.1 and 1.2

22. How many kilograms of ammonia are in a 3000 ft³ tank when the gauge pressure is 950 lb/in², and the ammonia temperature is 31°C?

    atomic weights of nitrogen and hydrogen = 14 and 1, respectively

$$NH_3 = 17 \text{ grams/gram-mole} \quad 3000 \text{ ft}^3 \times \frac{28.32 \text{ L}}{\text{ft}^3} = 84,960 \text{ L}$$

$$P_{absolute} = \frac{950 \text{ lb/in}^2 + 14.7 \text{ lb/in}^2}{14.7 \text{ lb/in}^2/\text{atmosphere}} = 65.6 \text{ atmospheres}$$

$$T = 31°C + 273 = 304 \text{ K}$$

$$n = \frac{PV}{RT} = \frac{65.6 \text{ atm} \times 84,690 \text{ L}}{(0.0821 \text{ L-atm/mole-K} \times 304 \text{ K})} = 223,579 \text{ gram-moles NH}_3$$

223,579 g-mols $\times$ 17 g/mol = 3,800,843 g = 3801 kg

Answer: 3801 kilograms of NH_3.

23. What would be the volume if the gas in problem 22 expanded to atmospheric pressure at a temperature of 20°C as might occur during a rapid tank rupture?

P, V, and T = pressure, volume, and absolute temperature of the gas, respectively

i and f = initial and final conditions, respectively

$$\frac{P_i V_i}{T_i} = \frac{P_f V_f}{T_f}$$

rearranging:

$$V_f = \frac{P_i T_f V_i}{P_f T_i} = 3000 \text{ ft}^3 \times \frac{65.6 \text{ atm}}{1 \text{ atm}} \times \frac{293°K}{304°K} = 189,679 \text{ ft}^3$$

Answer: 189,679 cubic feet.

24. The dust concentration in a limestone mill is 41 mppcf. The density of $CaCO_3$ is 2.71. If the calcite particles are spherical with a diameter of 1.42 microns, how much limestone dust is in the air of a 40,000 ft³ ball mill plant? How much dust is in every liter of mill air inhaled by the ball mill operators?

Φ = 1.42 microns = 0.000142 cm   radius = 0.000071 cm

$$\text{volume of sphere} = \frac{4}{3}\pi r^3 = \frac{4}{3}\pi (0.000071 \text{ cm})^3 = 1.5 \times 10^{-12} \text{ cm}^3/\text{dust particle}$$

$$40,000 \text{ ft}^3 \times \frac{1.5 \times 10^{-12} \text{ cm}^3}{\text{particle}} \times \frac{41 \times 10^6 \text{ particles}}{\text{ft}^3} = 2.46 \text{ cm}^3 \times 2.71 \text{g/cm}^3 =$$

6.67 grams = 6670 mg

40,000 ft³ $\times$ 28.32 L/ft³ = 1,132,800 L

6670 mg $CaCO_3$/1,132,800 L = 0.00589 mg/L = 5.89 mcg/L

Answers: 6670 milligrams. 5.89 micrograms of $CaCO_3$ dust/liter of air.

25. Air was sampled with a midget impinger for one hr and 17 min at an average rate of 0.89 L/m. How much ozone gas was present if the chemist detected 3.6 mcg O_3 per mL, the impinger collection efficiency was 71%, and there were 13.5 mL of potassium iodide ozone collection solution?

0.89 L/min $\times$ (60 + 17) min = 68.5 Lmol. wt. O_3 = 16 $\times$ 3 = 48

(100/71) = impinger inefficiency collection factor = 1.408

3.6 mcg/mL $\times$ 13.5 mL $\times$ 1.408 = 68.4 mcg O_3

$$ppm = \frac{\dfrac{mcg}{L} \times 24.45}{molecular\ weight} = \frac{\dfrac{68.4\ mcg}{68.5\ L} \times 24.45}{48} = 0.51\ ppm$$

Answer: 0.51 ppm ozone gas.

26. Convert 136 micrograms of ethyl alcohol vapor per liter into ppm (volume/volume).

molecular weight CH_3CH_2OH = 12 + 12 + 6 + 16 = 46

$$ppm = \frac{\dfrac{136\ mcg\ EtOH}{L} \times 24.45}{46} = 72.3\ ppm$$

Answer: 72 ppm EtOH vapor.

27. An analyst counts 3.4 fibers/field on an aerosol filter. There are 27,900 fields/filter. What was the fiber concentration in f/cc if air sampling was 2 L/min for 89 min?

3.4 fibers/field $\times$ 27,900 fields = 94,860 fibers

$$\frac{94,680\ fibers}{(2000\ cc/min) \times 89\ minutes} = 0.53\ f/cc = 530,000\ fibers/M^3$$

Answer: 0.53 fiber/cubic centimeter of air.

28. What is the concentration of nitrogen in air in ppm?

air: 78.09% N_2 + 20.95% O_2 + 0.93% argon + $\cong$ 0.03% CO_2 + trace gases ($\cong$ 79% "inerts" + $\cong$ 21% O_2)

Answer: 780,900 ppm N_2 (i.e., 100% air = 106 ppm).

29. The PEL for a metal is 0.2 mg/M^3. A chemist can reliably detect four micrograms with good accuracy and precision. At an air sampling rate of 1.1 L/m,

how long would an industrial hygienist have to sample the air to detect 10% of the PEL?

$PEL = 0.2 \text{ mg/M}^3 = 0.2 \text{ mcg/L}$ $10\% \text{ PEL} = 0.02 \text{ mcg/L}$

$\dfrac{4 \text{ mcg}}{0.02 \text{ mcg/L}} = $ at least 200 liters of air must be sampled

$\dfrac{200 \text{ L}}{1.1 \text{ L/min}} = $ at least 182 minutes

Answer: > 3 hrs.

30. What gas concentration results when five mL of dry ammonia gas are injected by a gas syringe into a 13 gallon glass calibration carboy of air?

method 1: $\dfrac{10^6 \times 5 \text{ mL}}{13 \text{ gallons} \times (3785 \text{ mL/gallon})} = 102 \text{ ppm NH}_3$

method 2:   17 g NH_3/24.45 L = 0.695 g/L = 0.695 mg/mL

0.695 mg/mL × 5 mL = 3.475 mg

13 gallons × 3.785 L/gallon = 49.205 L = 0.0492 M^3

$\dfrac{\dfrac{3.475 \text{ mg}}{0.0492 \text{ M}^3} \times 24.45}{17} = 102 \text{ ppm NH}_3$

method 3: $\dfrac{5 \text{ mL}}{49,205 \text{ mL}} = \dfrac{\times \text{ ppm}}{10^6 \text{ ppm}}$ $x = 102 \text{ ppm}$

Answer: The instrument calibration bottle contains 102 ppm NH_3 gas.

31. What is the gas concentration after 75 mL of pure CO gas mixes with air containing two ppm CO in a 313 L instrument calibration tank? Assume negligible dilution loss as the CO gas is quickly injected into the tank.

$\dfrac{10^6 \times 0.075 \text{ L}}{313 \text{ L} + 0.075 \text{ L}} = 240 \text{ ppm} = \left[\dfrac{75 \text{ mL}}{313,000 \text{ mL}}\right] \times 10^6 \ (240 + 2) \text{ ppm} = 242 \text{ ppm}$

Answer: Approximately 242 ppm carbon monoxide gas.

32. A chemist dropped a chlorine bottle releasing two pounds of gas into a laboratory with no ventilation. He immediately left and returned wearing a SCBA by which time the gas had mixed uniformly throughout the laboratory. The laboratory is 14 × 20 × 40 ft. He turned on an exhaust hood with a uniform face velocity of 170 ft/min. The hood face dimensions are 40 × 66 in. How long before the Cl_2

gas concentration is reduced to 0.2 ppm (20% of the 1 ppm STEL)? Assume ideal ventilation mixing.

$$\frac{40 \times 66 \text{ in}}{144 \text{ in}^2/\text{ft}^2} = 18.33 \text{ ft}^2 \times 170 \text{ fpm} = 3116 \text{ ft}^3/\text{minute}$$

$$\text{molecular weight } Cl_2 = 71$$

$$14 \times 20 \times 40 \text{ ft} = 11,200 \text{ ft}^3 = 317.3 \text{ M}^3 \quad 2 \text{ lb} = 908 \text{ g}$$

$$\frac{908 \text{ g}}{317.3 \text{ M}^3} = 2.86 \text{ g/M}^3 = 2860 \text{ mg/M}^3$$

$$\text{ppm} = \frac{\dfrac{2860 \text{ mg}}{\text{M}^3} \times 24.45}{71} = 985 \text{ ppm } Cl_2$$

$$t = \frac{-\ln\left[\dfrac{C}{C_0}\right]}{\dfrac{Q}{V}} = \frac{-\ln\left[\dfrac{0.2 \text{ ppm}}{985 \text{ ppm}}\right]}{\dfrac{3116 \text{ cfm}}{11,200 \text{ ft}^3}} = \frac{-\ln 0.000203}{0.2782/\text{min}} = \frac{-(-8.502)}{0.2782/\text{min}} = 30.6 \text{ min}$$

Answer: At least 31 min.

33. Twenty kilograms of methyl chloroform (molecular weight = 133.4) evaporate into a $10 \times 25 \times 35$-ft tank which has no ventilation. What is the equilibrium concentration in ppm? Is this a fire hazard? Could welding be permitted when the vapor level is reduced by ventilation to 50 ppm (1/7 of the TLV)? Could a worker enter wearing a half-mask twin charcoal cartridge respirator without special ventilation and taking other precautions?

$$20 \text{ kg} = 2 \times 10^7 \text{ mg} \quad 10 \times 25 \times 35 \text{ ft} = 8750 \text{ ft}^3 = 247.8 \text{ M}^3$$

$$\frac{2 \times 10^7 \text{ mg}}{247.8 \text{ M}^3} = 80,710 \text{ mg/M}^3$$

$$\text{ppm} = \frac{\dfrac{80,710 \text{ mg}}{\text{M}^3} \times 24.45}{133.4} = 14,793 \text{ ppm} = 1.48\% \text{ (vol/vol)}$$

Answers: 14,800 ppm. No, under normal conditions, this is not an explosion hazard. However, at very high concentrations in air and, especially, in high oxygen levels, an explosion could occur with high temperature ignition sources. $COCl_2$, HCl, dichloroacetylene, etc. could be generated in the welding arc. No, this is the wrong respirator since the maximum use conditions are 1000 ppm organic vapors for charcoal filters provided that the concentration immediately dangerous to life and health (IDLH) is not less than 1000 ppm. In other words, since the IDLH concentration for methyl chloroform is

700 ppm, this is the maximum concentration permitted for use with organic vapor charcoal cartridge respirators, not 1000 ppm. This would be an acceptable respirator if the vapor levels did not exceed 700 ppm.

34. Ten mL of dry ammonia gas are injected into a 152 L tank of pure air. What is the NH_3 gas concentration? Could most people detect the resultant ammonia gas concentration by smell or irritation?

molecular weight NH_3 = 17   density at NTP = 17g/24.45L = 0.7 mg/mL

10 mL NH_3 = 7 g $\dfrac{7\,g}{152\,L}$ = 0.046 mg/L = 46 mcg/L

ppm NH_3 = $\dfrac{\dfrac{46\ mcg}{L}\times 24.45}{17}$ = 66.2 ppm NH_3

Answers: 66 ppm NH_3. Yes, everyone would respond to this gas concentration.

35. If the apparent volume of sampled air is 570 L at 645 mm Hg and 33°C, what was the standard air volume that was sampled?

33°C + 273 = 306 K

liters = $\dfrac{298\times V\times P}{760\times T}$ = $\dfrac{(298\ K)(570\ L)(645\ mm\ Hg)}{(760\ mm\ Hg)(306\,K)}$ = 471 L

Answer: 471 L.

36. The LEL for a solvent vapor is 1.7%. What is the vapor concentration in ppm if a calibrated CGI indicates a reading of 64% of the LEL?

1.7% = 17,000 ppm.   0.64 LEL = 0.64 × 17,000 ppm = 10,880 ppm.

Answer: 10,880 ppm.

37. An impinger contains 13 mL of dilute alkali. Each mL can neutralize 0.012 mg of hydrochloric acid gas. Air was sampled at 0.86 L/m for 14.7 min when the alkali was neutralized as indicated by an abrupt color change in the solution. What was the average concentration of acid gas during the sampling period?

13 mL × 0.012 mg/mL = 0.156 mg HCl

molecular weight HCl   = 1 + 35.5 = 36.5

0.86 L/m × 14.7 min   = 12.64 L

$$ppm = \frac{\dfrac{156 \text{ mcg}}{12.64 \text{ L}} \times 24.45}{36.5} = 8.27 \text{ ppm HCl gas}$$

Answer: 8.3 ppm HCl, assuming 100% collection efficiency by the impinger.

38. 0.1 mL of a solvent mixture is evaporated in a 313 L calibration tank. The mixture is comprised (by volume) of 30% MEK, 30% toluene, and 40% methylene chloride with respective densities of 0.805, 0.870, and 1.335 g/mL. Calculate the ppm of each vapor and the apparent molecular weight of the mixture. Molecular weights = 72, 92, and 85, respectively.

MEK: 0.03 mL × 0.805 g/mL = 0.02415 g
toluene: 0.03 mL × 0.870 g/mL = 0.0261 g
CH_2Cl_2:   0.04 mL × 1.335 g/mL = 0.0534 g
           0.10 mL   0.1037 g

$$ppm \text{ MEK} = \frac{\dfrac{24{,}150 \text{ mcg}}{313 \text{ L}} \times 24.45}{72} = 26 \text{ ppm}$$

$$ppm \text{ toluene} = \frac{\dfrac{26{,}100 \text{ mcg}}{313 \text{ L}} \times 24.45}{92} = 22 \text{ ppm}$$

$$ppm \text{ } CH_2Cl_2 = \frac{\dfrac{53{,}400 \text{ mcg}}{313 \text{ L}} \times 24.45}{85} = 49 \text{ ppm}$$

26 ppm + 22 ppm + 49 ppm = 97 ppm total solvent vapors

0.1037 g/313 L = 331 mcg/L

$$\text{apparent molecular weight} = \frac{\dfrac{331 \text{ mcg}}{L} \times 24.45}{97 \text{ ppm}} = 83.4$$

Answers: 26 ppm MEK, 22 ppm toluene, and 49 ppm CH_2Cl_2. The apparent molecular weight = 83.4.

39. How many molecules of TDI are in every two-liter inhalation if the air contains 0.001 ppm (one ppb) TDI vapor?

$$\frac{6.023 \times 10^{23} \text{ molecules TDI/gram-mole}}{24.45 \text{ L/gram-mole}} = 10^6 \text{ ppm} = 100\%$$

$$1 \text{ ppb} = \frac{10^{-9} \times 6.023 \times 10^{23} \text{molecules TDI}}{24.45 \text{ L}} =$$

$$\frac{2.46 \times 10^{13}}{\text{L}} \times 2 \text{ liters} = 4.92 \times 10^{13} \text{ molecules}$$

Answer: $\cong 5 \times 10^{13}$ molecules of TDI in every two-liter inhalation.

40. A rotameter calibrated at 25°C and 760 mm Hg indicates a rate of 1.45 L/min at 33°C and 690 mm Hg. What is the corrected standard air flow rate?

$$\text{L/min} = 1.45 \text{ L/min} \times \sqrt{\frac{690 \text{ mm Hg}}{760 \text{ mm Hg}}} \times \sqrt{\frac{25°\text{C} + 273 \text{ K}}{33°\text{C} + 273 \text{ K}}} =$$

$1.45 \times 0.9528 \times 0.9868 = 1.36$ L/min

Answer: 1.28 L of air/min (i.e., the slightly colder and denser air at standard conditions has greater buoyancy for the rotameter float ball). The square root function is used for rotameters and critical orifice air flow meters, but not for dry and wet gas meters. Refer to problem 311.

41. Give two basic industrial hygiene air sampling "rules of thumb."

Answers:

1. minimum collection volume (in M³) = $\dfrac{\text{analytical sensitivity (mg)}}{0.1 \times \text{TLV} \left(\text{in mg/M}^3\right)}$

2. minimum sampling time (in hours) $\dfrac{\text{analytical sensitivity (mg)}}{\left(0.1 \times \text{TLV in mg/M}^3\right) \times \left(\text{rate in M}^3/\text{hour}\right)}$

Generally, an air sample should be taken to detect at least 10% of its respective Threshold Limit Value, permissible exposure limit, short-term exposure limit, ceiling limit, Workplace Environmental Exposure Limit, or NIOSH-recommended exposure limit.

42. What is the error of measurement if the true value is 13 ppm and the amount found (the experimental value) is 16 ppm?

$$\% \text{ error} = \frac{\text{EV} - \text{TV}}{\text{TV}} \times 100 = \frac{16 \text{ ppm} - 13 \text{ ppm}}{13 \text{ ppm}} \times 100 = 23\%$$

Answer: + 23%.

43. A bottle containing 600 g of titanium tetrachloride breaks in a warehouse. The liquid quickly evaporates hydrolyzing in highly humid air according to the reaction:

$$TiCl_4 + 2 H_2O \rightarrow TiO_2 \uparrow + 4 HCl \uparrow$$

The warehouse (20 × 60 × 185 ft) is empty, unventilated, and unoccupied. Is it safe to enter without wearing respiratory protection? What concentration of HCl gas and TiO_2 fumes are present? Molecular weight of $TiO_2 = 47.09$. Assume sufficient water vapor as atmospheric moisture to provide stoichiometric conversion.

20 × 60 × 185 ft = 222,000 ft³ = 6289 M³

$$\frac{600 \text{ g TiCl}_4}{189.73 \text{ g/mole}} = 3.16 \text{ moles TiCl}_4$$

therefore: molecular weight

2 × 3.16 mols H_2O required   18
1 × 3.16 mols TiO_2 produced   79.9
4 × 3.16 mols HCl produced   36.5

1 × 3.16 mols TiO_2 × 79.2 g/mol = 252,484 mg TiO_2

4 × 3.16 mols HCl × 36.5 g/mol = 461,360 mg HCl

$$\frac{252,484 \text{ mg TiO}_2}{6289 \text{ M}^3} = 40.1 \text{ mg TiO}_2 \text{ fume/M}^3$$

$$\frac{\frac{461,360 \text{ mg HCl}}{6289 \text{ M}^3} \times 24.45}{36.5} = 49.1 \text{ ppm HCl gas}$$

Answers: 40 mg TiO_2/M³. 49 ppm HCl. OSHA PEL for HCl = five ppm ceiling. No, *Stay Out!* Ventilate. Otherwise, enter with a SCBA or full-face acid gas cartridge respirator with HEPA pre-filters. A SCBA is preferred since the IDLH for HCl is 50 ppm, barely above the average concentration of HCl gas in the room.

44. How much liquid toluene is needed to make 100 ppm of vapor in a 10 × 14 × 20-ft chamber when the atmospheric pressure is 640 mm Hg and the air temperature is 21°C?   10 × 14 × 20 ft = 2800 ft³

$$mL = \frac{100 \text{ ppm x} \frac{92 \text{ g}}{\text{g-mole}} \times \frac{2800 \text{ ft}^3}{35.3 \text{ ft}^3/\text{M}^3} \times 273 \text{ K} \times 640 \text{ mm Hg}}{\frac{0.867 \text{ g}}{\text{mL}} \times \frac{22.4 \text{ L}}{\text{g-mole}} \times 294 \text{ K} \times 760 \text{ mm Hg} \times 10^3}$$

Answer: 29.4 mL.

45. What concentration of solvent vapor remains in a 1000 gallon tank (containing no liquid solvent) after 500 gallons of the air: vapor mixture have been removed and replaced with clean air? The initial concentration was 1000 ppm solvent vapor.

$$(2.3 \log 1000 \text{ ppm}) - (2.3 \log \text{y ppm}) = \frac{500 \text{ gallons}}{1000 \text{ gallons}}$$

$$(2.3 \times 3) - (2.3 \log \text{y ppm}) = 0.5 \qquad 2.3 \log \text{y ppm} = 6.4$$

$$\log \text{y ppm} = \frac{6.4}{2.3} = 2.783 \qquad \text{y} = 606 \text{ ppm}$$

Answer: 606 ppm.

46. Calculate the vapor volume of one g of water when boiled at 730 mm Hg.

$$P V = n R T \qquad n = 1 \text{ g}/18 \text{ g/mol} = 0.0556 \text{ mol}$$

$$T = 100°C + 273 \text{ K} = 373 \text{ K}$$

$$V = \frac{(0.0556 \text{ mole})(0.082)(373 \text{ K})}{\dfrac{730 \text{ mm Hg}}{760 \text{ mm Hg}}} = 1.7705 \text{ L}$$

Answer: 1771 mL of water vapor, i.e., one mL of liquid water produces 1.77 L of steam under these conditions of temperature and pressure.

47. A stack gas sample was collected using a dry gas meter calibrated at 32°F and 760 mm Hg. The water vapor pressure at this temperature (absolute humidity) is 0.08 lb/in². The wet and dry bulb temperatures of the stack gas were 96 and 111°F, respectively (58% relative humidity). Corresponding water vapor pressure is 0.78 lb/in². The indicated air volume was 930 ft³. The barometric pressure at the time of sampling was 740 mm hg. What is the corrected dry gas volume?

   Stack gas volumes are often calculated as if the gases are dry since the variable water vapor, especially at high temperatures, can account for a significant portion of the gas volume. The variations due to pressure, temperature, and water vapor content are calculated by:

$$V_1 = \frac{V_2(P_2 - W_2)(273 \text{ K} + T_1)}{(P_2 - W_1)(273 \text{ K} + T_2)}, \text{ where:}$$

V_2 = apparent gas volume at T_2 (°C)

V_1 = calculated gas volume at T_1 (°C)

W_1 and W_2 = mm H_2O vapor pressure at calculated and observed conditions, respectively

P_1 and P_2 = calculated and observed barometric pressures, respectively

$(lb/in^2) \times 51.71$ = mm Hg

0.02 lb/in^2 = 1.03 mm Hg   32°F = 0°C

0.78 lb/in^2 = 40.33 mm Hg   112°F = 44.4°C

$$V_1 = \frac{(930 \text{ ft}^3)(740 \text{ mm Hg} - 40.3 \text{ mm Hg})(273 \text{ K})}{(760 \text{ mm Hg} - 1 \text{ mm Hg})(273 \text{ K} + 44.4°C)} = 737.4 \text{ ft}^3$$

Answer: 737.4 cubic feet of dry gas.

48. 1000 L of dry nitrogen gas at one atmosphere pressure and 20°C are adiabatically compressed to 5% of the initial gas volume. What is the final temperature and pressure? For N_2 gas, $\alpha = 1.4$ (the specific heat. Refer to problem 21.).

$P_1 (V_1)^\alpha = P_2 (V_2)^\alpha$

(one atm.) $(V_1)^{1.4} = P_2 (V_1/20)^{1.4}$

$$P_2 = \frac{(1 \text{ atm})(V_1)^{1.4}}{(V_1/20)^{1.4}} = \frac{(1 \text{ atm})(1000 \text{ L})^{1.4}}{(50 \text{ L})^{1.4}} = 66.3 \text{ atmospheres}$$

$$T_1(V_1)^{\alpha-1} = T_2(V_2)^{\alpha-1} = (293 \text{ K})(1000 \text{ L})^{1.4-1} = T_2(50 \text{ L})^{1.4-1}$$

$$T_2 = \frac{293 \text{ K}(1000 \text{ L})^{1.4-1}}{(50 \text{ L})^{1.4-1}} = \frac{293 \text{ K}(1000 \text{ L})^{0.4}}{(50 \text{ L})^{0.4}} = 970.95 \text{ K}$$

Answers: 66.3 atmospheres. 971 K.

49. An air sample filter contained 36 micrograms of chromium. If all of the chromium was from a lead chromate paint spray aerosol ("school bus yellow"), how much lead chromate was present?

$PbCrO_4$ molecular weight = 207 + 52 + (4 × 16) = 323

323/51 = 6.21   36 mcg Cr × 6.21 = 224 mcg $PbCrO_4$

Answer: 224 micrograms $PbCrO_4$.

50. Assume a worker has an average inhalation of 15 L/min and inhaled air contains nine mcg cobalt/M^3. If his absorption is 25%, how much Co would the worker accumulate every 8-hr work shift? Disregard excretion during exposure period.

(9 mcg Co/M³) × 0.25 = 2.25 mcg Co/M³ = 0.00225 mcg/L

$$\frac{0.00225 \text{ mcg Co}}{L} \times \frac{15 \text{ L}}{\text{minute}} \times \frac{60 \text{ minutes}}{\text{hour}} \times \frac{8 \text{ hours}}{\text{shift}} = 16.2 \text{ mcg Co/shift}$$

Answer: 16 micrograms of cobalt per eight-hr work shift.

51. Air temperature = 42°C. Barometric pressure = 718 mm Hg. Air sampling rate = 2.31 L/m. Sampling time = 17.5 min. Assume the air is saturated with water vapor (61.5 mm Hg). What is the sampled volume of dry air at 25°C and 760 mm Hg?

17.5 min × 2.31 L/m = 40.425 L

$$40.425 \text{ L} \times \frac{718 \text{ mm Hg} - 61.5 \text{ mm Hg}}{760 \text{ mm Hg}} \times \frac{298 \text{ K}}{273 \text{ K} + 42°C} = 33 \text{ L}$$

Answer: 33 L of dry air.

52. What is the mercury vapor concentration in a dynamic vapor generation system if 100 mL/min saturated at 18.8 mcg Hg/L are diluted with mercury-free air at 27 L/min?

$$C_t = \frac{A \times C}{A + B} = \text{ concentration of contaminant in dynamic generation calibration}$$
system

where:

A = contaminant flow rate
B = clean air flow rate
C = concentration of contaminant in A

100 mL/min = 0.1 L/min

$$C_t = \frac{(0.1 \text{ L/min}) \times (18.8 \text{ mcg/L})}{(27 \text{ L/min}) + (0.1 \text{ L/min})} = 0.069 \text{ mcg/L}$$

Answer: 0.069 mg Hg vapor/M³.

53. How many particles result from crushing one cubic centimeter of quartz into one micron cubic particles?

1 cc = 1 cm³   1 meter = 10⁶ microns
1 cm = 10⁴ microns   (10⁴ microns)³ = 10¹² particles

Answer: 10¹² particles of respirable dust, i.e., 1,000,000,000,000 particles.

54. What is the saturation concentration (in mg/M³) of mercury vapor at 147°F? Hg vapor pressure at this temperature is 0.0328 mm Hg. Assume the air temperature is also 147°F. Molecular weight of Hg = 200.6 g/g-mol.

$$\frac{0.0328 \text{ mm Hg}}{760 \text{ mm Hg}} \times 10^6 = 43.16 \text{ ppm} \qquad 147°F = 63.89°C$$

$$22.4 \text{ L} \times \frac{273 \text{ K} + 63.89°C}{273 \text{ K}} = 27.64 \text{ L} \qquad \frac{mg}{M^3} = \frac{43.16 \text{ ppm} \times 200.6}{27.64 \text{ L}} = \frac{313 \text{ mg}}{M^3}$$

Answer: 313 mg mercury/M³ (OSHA PEL = 0.05 mg/M³).

55. How much mercury would have to evaporate to produce a concentration of 0.1 mg/M³ in a chloralkali plant with a 2,250,000 ft³ interior volume? Density of liquid mercury = 13.6 g/mL.

0.1 mg/M³ = 2.83 mcg Hg/ft³   2.83 mcg/ft³ × 2.25 × 10⁶ ft³ = 6.37 × 10⁶ mcg

$$\frac{6.37 \text{ g Hg}}{13.6 \text{ g/mL}} = 0.47 \text{ mL}$$

Answers: 6.4 g.   0.47 milliliter.

56. If one cc of air has a mass of 1.2 mg at 25°C and 760 mm Hg, what is the density of mercury vapor? Molecular weight of Hg = 200.6.

$$1 \text{ ppm Hg vapor} = \frac{200.6 \times mg/L}{24,450 \text{ mL/g-mole}} = 0.0082 \text{ mg/L} = 8.2 \text{ mcg/L} =$$

8.2 /1000 cc = 0.0082 g/cc = 8.2 mg/cc

8.2 mg/M³ = 1 cc/M³   1 cc Hg vapor = 8.2 mg

$$1 \text{ cc air} = 1.2 \text{ mg} \qquad \frac{8.2 \text{ mg Hg/cc}}{1.2 \text{ mg air/cc}} = 6.83$$

Alternatively, one could use the ratio of their molecular weights, i.e., the "apparent" molecular of air = 28.94, or 200.6/28.94 = 6.93 — essentially identical to the above. (See problem 58.)

Answer: 6.83 (air = 1.00, no units).

57. How many pounds of air are inhaled weekly by a person with a daily inhalation of 22.8 M³, the air inhalation volume of "standard man" (70 kg)?

$$\frac{22.8 \text{ M}^3}{day} \times \frac{7 \text{ days}}{week} \times \frac{35.3 \text{ ft}^3}{M^3} \times \frac{0.075 \text{ lb}}{ft^3} = \frac{423 \text{ lb}}{week}$$

Answer: About 400–450 pounds per week for a 154-pound man or woman who has an average daily metabolic rate. The inhaled volume, of course, increases as metabolic activity increases and as the person's weight increases.

58. Calculate dry air's "apparent" molecular weight if the atomic weight of argon = 39.9.

	% by volume		molecular weight		proportion of molecular weight
O_2	21.0	×	32	=	6.72
N_2	78.1	×	28	=	21.866
Ar	0.9	×	39.9	=	0.355
	100.0				28.941

Answer: The apparent molecular weight of dry air = 28.94.

59. Worker breathing zone TWAE air contains 30 mcg Pb/M³ (PEL = 50 mcg/M³) and 0.8 mg H_2SO_4 mist/M³ (PEL = 1 mg/M³). Is the PEL for the mixture exceeded?

$$\frac{30 \text{ mcg/M}^3}{50 \text{ mcg/M}^3} = 0.6, \text{ or } 60\% \text{ of PEL} \qquad \frac{0.8 \text{ mg H}_2\text{SO}_4/\text{M}^3}{1 \text{ mg H}_2\text{SO}_4/\text{M}^3} = 0.8, \text{ or } 80\% \text{ of PEL}$$

Answer: No, the toxic effects of these air contaminants are independent of each other. However, since both exceed their respective action levels, an aggressive industrial hygiene control program is indicated.

60. By how much should workers' exposure limits be reduced if they are exposed for a 12-hr work shift?

$$\text{exposure reduction factor} = \frac{8}{h} \times \frac{24 - h}{16}, \text{ where } h = \text{hours exposed per work day}$$

$$\frac{8}{12} \times \frac{24 - 12}{16} = 0.5$$

Answer: By at least 50%. This accounts, in part, for the reduced time available each day to detoxify and reverse the daily effects.

61. A liquid contains (by weight) 50% heptane (TLV = 400 ppm = 1600 mg/M³), 30% methyl chloroform (TLV = 350 ppm = 1900 mg/M³), and 20% perchloroethylene (TLV = 50 ppm = 335 mg/M³). Assume complete evaporation of all solvents in the mixture. What is the TLV for the vapor mixture?

n-heptane:	1 mg/M³	$\cong 0.25$ ppm
CH_3CCl_3:	1 mg/M³	$\cong 0.18$ ppm
"perc":	1 mg/M³	$\cong 0.15$ ppm

$$\text{TLV of mixture} = \frac{1}{\dfrac{0.5}{1600} + \dfrac{0.3}{1900} + \dfrac{0.2}{335}} = 935 \text{ mg/M}^3 \quad \text{Of this mixture,}$$

$$n\text{-heptane:} \quad 935 \text{ mg/M}^3(0.5) = \frac{468 \text{ mg}}{\text{M}^3} \times 0.25 = 117 \text{ ppm}$$

$$CH_3CCl_3: \quad 935 \text{ mg/M}^3(0.3) = \frac{281 \text{ mg}}{\text{M}^3} \times 0.18 = 51 \text{ ppm}$$

$$\text{"perc":} \quad 935 \text{ mg/M}^3 (0.2) = \frac{187 \text{ mg}}{\text{M}^3} \times 0.15 = 29 \text{ ppm}$$

Answer: 117 ppm + 51 ppm + 29 ppm = 197 ppm = 935 mg/M³.

62. A process evaporates 5.7 pounds of isopropyl alcohol of vapor per hour into a work area which has a general ventilation rate of 4500 cfm. What is the average steady state vapor concentration?

$$\text{ppm} = \frac{\text{ER} \times 24.45 \times 10^6}{Q \times \text{molecular weight}}, \text{ where:}$$

ER = evaporation (generation) rate (in g/min), and

Q = ventilation rate (in L/min)

$CH_3CHOHCH_3$ molecular weight = C_3H_8O = 36 + 8 + 16 = 60

$$\frac{5.7 \text{ lb}}{\text{hour}} \times \frac{454 \text{ g}}{\text{lb}} \times \frac{\text{hour}}{60 \text{ minutes}} = \frac{43.13 \text{ grams}}{\text{minutes}}$$

$$Q = \frac{4500 \text{ ft}^3}{\text{minute}} \times \frac{28.3 \text{ L}}{\text{ft}^3} = 127,350 \text{ L/m}$$

$$\text{ppm} = \frac{\dfrac{43.13 \text{ grams}}{\text{minute}} \times 24.45 \times 10^6}{\dfrac{127,350 \text{ L}}{\text{minute}} \times 60} = 138 \text{ ppm IPA}$$

Answer: 138 ppm isopropyl alcohol vapor.

63. The dry bulb temperature in Miami Beach is 80°F at a barometric pressure of 760 mm Hg and with a relative humidity of 40%. The water vapor pressure at these conditions is 0.195 lb/in². A hurricane is approaching. What is the concentration of H_2O vapor in the air as the barometer decreases from 760 to 680 mm Hg?

(lb/in²) × 51.71 = mm Hg   (0.195 lb/in²) × 51.71 = 10.08 mm Hg

$$\frac{10.08 \text{ mm Hg}}{680 \text{ mm Hg}} \times 10^6 = 14,824 \text{ ppm } H_2O \text{ vapor}$$

$$ppm = \frac{mg}{L} \times \frac{22,400}{18} \times \frac{299.67 \text{ K}}{273 \text{ K}} \times \frac{760 \text{ mm Hg}}{680 \text{ mm Hg}},$$

or:

$$\frac{mg \, H_2O}{L} = \frac{14,824 \text{ ppm}}{\dfrac{22,400}{18} \times \dfrac{299.67 \text{ K}}{273 \text{ K}} \times \dfrac{760 \text{ mm Hg}}{680 \text{ mm Hg}}} = \frac{9.708 \text{ mg}}{L}$$

Answers: 14,824 ppm ($\cong$ 1.48%) $\cong$ 9,700 mg H_2O/M^3. At the former conditions, before the barometric pressure dropped, the water vapor concentration was:

$$\frac{10.08 \text{ mg Hg}}{760 \text{ mg Hg}} \times 10^6 = 13,263 \text{ ppm } H_2O \text{ vapor.}$$

The increase in the water vapor concentration in the atmosphere as the barometric pressure decreases explains, in part, why heavy rains accompany hurricanes.

64. During an earthquake, two adjacent compressed gas lines in a 20 × 125 × 300-ft unventilated, closed room in a chemical plant simultaneously burst. One pipeline released 20 pounds of anhydrous ammonia, and the other released four pounds of anhydrous hydrogen chloride. Dense white fume and a pungent, highly irritating gas immediately formed. Assume complete reaction of both gases. What was the white fume? What was the remaining gas and its concentration? Should we boldly stroll into this room without respiratory protection? If not, what type of respirator should we use?

$$HCl + NH_3 \rightarrow NH_4Cl \uparrow$$

$$4 \text{ lb HCl} + 20 \text{ lb } NH_3 \rightarrow \times \text{ lb } NH_4Cl = 1816 \text{ g HCl} + 9080 \text{ g } NH_3 \rightarrow y \text{ g } NH_4Cl$$

$$\frac{1816 \text{ g HCl}}{36.5 \text{ g HCl/g-mole}} = 49.75 \text{ moles HCl}$$

$$\frac{9080 \text{ g } NH_3}{17 \text{ g } NH_3/\text{g-mole}} = 534.1 \text{ moles } NH_3$$

Therefore, from the above equation, stoichiometrically, only 49.75 mols of NH_4Cl can be formed (molecular weight NH_4Cl = 53.5 g/g-mol):

$$\text{y grams NH}_4\text{Cl} = 49.75 \text{ moles} \times \frac{53.5 \text{ grams}}{\text{mole NH}_4\text{Cl}} = y$$

$$= 2662 \text{ grams NH}_4\text{Cl} = 5.86 \text{ lb}$$

$$20 \times 125 \times 300 \text{ ft} = 750,000 \text{ ft}^3 = 21,247 \text{ M}^3$$

$$\frac{2262 \text{ g}}{22,247 \text{ M}^3} = \frac{0.125 \text{ g}}{\text{M}^3} = \frac{125 \text{ mg NH}_4\text{Cl fume}}{\text{M}^3}$$

20 lb NH_3 (an excess) + 4 lb HCl (which totally reacts) → 5.86 lb NH_4Cl fume

24 lb total reactants – 5.86 lb product = 18.14 lb NH_3 gas remaining = 8236 g NH_3

$$\text{ppm NH}_3 = \frac{\dfrac{8,236,000 \text{ mg}}{21,247 \text{ M}^3} \times 24.45 \text{ L/gram-mole}}{17 \text{ grams NH}_3/\text{gram-mole}} = 557 \text{ ppm NH}_3$$

Answers: NH_4Cl. 2660 g, or 125 mg/M³. 557 ppm NH_3. No. SCBAs. The IDLH for ammonia is 300 ppm.

65. What is the eight-hr TWAE PEL for an insecticide mixture containing one part by weight "Parathion" (PEL = 0.1 mg/M³) and two parts "EPN" by weight (PEL = 0.5 mg/M³)?

$$\frac{C_1}{0.1 \text{ mg/M}^3} + \frac{C_2}{0.5 \text{ mg/M}^3} = \frac{C_{\text{mixture}}}{T_{\text{mixture}}} = \frac{C_m}{T_m} \quad C_2 = 2\,C_1 \quad C_m = 3\,C_1$$

$$\frac{C_1}{0.1 \text{ mg/M}^3} + \frac{2\,C_1}{0.5 \text{ mg/M}^3} = \frac{3\,C_1}{T_m} \qquad \frac{7\,C_1}{0.5 \text{ mg/M}^3} = \frac{3\,C_1}{T_m}$$

$$T_m = \frac{1.5}{7} = 0.21 \text{ mg/M}^3$$

Answer: PEL for the insecticide mixture = 0.21 mg/M³.

66. Calculate the TLV for a mineral dust mixture of 40% X (TLV = 1 mg/M³) and 60% Y (TLV = 0.3 mg/M³). The adverse effects on respiratory health are assumed to be additive (pulmonary fibrosis).

$$\frac{C}{\text{TLV}} = \frac{0.4}{1} + \frac{0.6}{0.3} \qquad \frac{1}{\text{TLV}} = 0.4 + 2.0 = 2.4$$

1 = 2.4 × the TLV   TLV = 1/2.4 = 0.42 mg/M³

Answer: $\text{TLV}_{x \text{ and } y}$ = 0.42 mg/M³.

67. Air contains 234 ppm acetone (TLV = 750 ppm), 119 ppm *sec*-butyl acetate (TLV = 200 ppm), 113 ppm MEK (TLV = 200 ppm), and 49 ppm methyl chloroform (TLV = 350 ppm). What is the concentration of the vapor mixture? Is the TLV exceeded?

234 ppm + 119 ppm + 133 ppm + 49 ppm = 535 ppm

$$\frac{234 \text{ ppm}}{750 \text{ ppm}} + \frac{119 \text{ ppm}}{200 \text{ ppm}} + \frac{133 \text{ ppm}}{200 \text{ ppm}} + \frac{49 \text{ ppm}}{350 \text{ ppm}} = 1.71$$

Answers: 535 ppm. Yes, by 71%.

68. What is the partial pressure of O_2 in dry air at sea level?

oxygen: 20.95 volume % in air 0.2095×760 mm Hg = 159.2 mm Hg

Answer: 159.2 mm Hg.

69. If, in problem 68, the barometric pressure does not change, but the air is humidified to 100% relative humidity at 25°C, how will the air composition be altered? Water vapor pressure at 25°C = 23.8 mm Hg.

O_2:	20.95% (760 mm Hg – 23.8 mm Hg) =	154.23 mm Hg
N_2, etc.:	79.05% (760 mm Hg – 23.8 mm Hg) =	581.97 mm Hg
H_2O:	=	23.8 mm Hg
		760 mm Hg

Answers: O_2 = 154.2 mm Hg. N_2 + argon + etc. trace gases = 582 mm Hg.

70. What is the percent oxygen in air at 12,000 ft altitude?

Answer: Although the partial pressure of O_2 decreases with altitude, the percent composition of air does not. The atmosphere remains $\cong$ 21% O_2 at any altitude.

71. If an air sample humidified to 50% at 25°C is taken from sea level to an altitude with a barometric pressure of 600 mm Hg and the same temperature, what will be the partial pressures of air gases and water vapor?

If 100% relative humidity at 25°C = 23.8 mm Hg, then 50% relative humidity at 25°C = 11.9 mm Hg.

$$H_2O \text{ vapor} = \frac{600 \text{ mm Hg}}{760 \text{ mm Hg}} \times 11.9 \text{ mm Hg} = 9.39 \text{ mm Hg}$$

O_2:	20.95% (600 mm Hg – 9.39 mm Hg)	= 123.73 mm Hg
N_2, etc.:	79.05% (600 mm Hg – 9.39 mm Hg)	= 466.88 mm Hg
		$\cong$ 600.00 mm Hg

Answers: H_2O = 9.4 mm Hg. O_2 = 123.7 mm Hg. N_2, etc. = 466.9 mm Hg.

72. An air filter is used at 1.36 L/m for 26 min to collect a monodisperse aerosol of uniformly spherical one-micron particles with a density of 2.6. If the concentration of particles is 7.8 mppcf, how many particles are collected on the filter? How much mass is collected? What is the dust concentration in mg/M³?

$$\frac{1.36\,\text{L}}{\text{minute}} \times 26 \text{ minutes} = 35.36\,\text{L} = 1.25\,\text{ft}^3$$

7.8 mppcf $\times$ 1.25 ft³ = 9.75 $\times$ 10⁶ particles

d = 1.0 micron = 0.0001 cm   r = 0.00005 cm

$$V = \frac{4}{3}\,\pi\,r^3 = \frac{4}{3}\,\pi\,(0.00005\,\text{cm})^3 = \frac{5.236 \times 10^{-13}\,\text{cm}^3}{\text{particle}}$$

$$\frac{5.236 \times 10^{-13}\,\text{cm}^3}{\text{particle}} \times 9.75 \times 10^6 \text{ particles} = 5.11 \times 10^{-6}\,\text{cm}^3$$

$$\frac{5.11 \times 10^{-6}\,\text{cm}^3}{\text{total particles}} \times \frac{2600\,\text{mg}}{\text{cm}^3} = 0.0133\,\text{mg} \qquad \frac{13.3\,\text{mcg}}{35.36\,\text{L}} = \frac{0.376\,\text{mg}}{\text{M}^3}$$

Answers: 9.8 $\times$ 10⁶ particles. 0.376 mg/M³.

73. Assuming a normal probability distribution of dust particles suspended in air, what is the standard geometric deviation if the geometric mean = 1.25 microns, and the 84.13% size = 2 microns?

$$\frac{2.0\,\text{microns}}{1.25\,\text{microns}} = 1.6\,\text{GSD}$$

Answer: Geometric standard deviation = 1.6.

74. 1.3 mL of nitrogen dioxide gas/min are diluted into an air stream of 5.7 ft³ per min. What is the NO₂ concentration in the mixed gas stream leading to the direct-reading instrument which is being calibrated?

$$\text{ppm} = \frac{\text{volume of gas/minute}}{\text{volume of air/minute}} \times 10^6 =$$

$$\frac{\dfrac{1.3\,\text{mL}}{\text{minute}}}{\dfrac{5.7\,\text{ft}^3}{\text{minute}} \times \dfrac{28.3\,\text{L}}{\text{ft}^3} \times \dfrac{1000\,\text{mL}}{\text{L}}} \times 10^6 = 8.06\,\text{ppm NO}_2$$

Answer: 8.1 ppm NO₂.

75. An industrial hygienist calibrates his stopwatch by telephoning the "Time Lady." He starts the watch when she says "At the tone, the time is 2:13 and 20 seconds." He stops the watch after calling her again when she says "2:27 and 10 sec." The watch's elapsed time is 13 min and 33 sec. What is the percent error of his watch?

$$2:27:10 = 2:26:70 \qquad 2:26:70 - 2:13:20 = 0:13:50$$

$$\% \text{ error} = \frac{\text{experimental value} - \text{true value}}{\text{true value}} \times 100 =$$

$$\frac{813 \text{ seconds} - 830 \text{ seconds}}{830 \text{ seconds}} \times 100 = -2.05\%$$

Answers: -2.05%. Watch correction factor $= \dfrac{830 \text{ seconds}}{813 \text{ seconds}} = 1.021$.

76. An aqueous solution of 250 ppm (w/v) lead nitrate is atomized into respirable mist droplets producing a total mist concentration of 360 mg/M³ (including water). What is the Pb concentration in air?

250 ppm $Pb(NO_3)_2$ = 250 mg $Pb(NO_3)_2$ /liter

molecular weight of $Pb(NO_3)_2 = 207 + (2 \times 14) + (6 \times 16) = 331$

$$\frac{207}{331} \times 100 = 62.5\% \text{ Pb} \qquad \frac{250 \text{ mg } Pb(NO_3)_2}{L} \times 0.625 = \frac{156 \text{ mg Pb}}{\text{liter}}$$

$$156 \text{ ppm} = 1.56 \times 10^{-4} = 0.000156 \qquad \frac{360 \text{ mg}}{M^3} \times 0.000156 = \frac{0.056 \text{ mg Pb}}{M^3}$$

Answer: 0.056 mg Pb/M³.

77. Total airborne particulates are emitted from an industrial power plant stack at a rate of 0.64 ton per day. What is the milligram/min emission rate?

$$\frac{0.64 \text{ ton}}{\text{day}} \times \frac{2000 \text{ pounds}}{\text{ton}} \times \frac{454 \text{ grams}}{\text{pound}} = \frac{581,120 \text{ grams}}{\text{day}}$$

$$\frac{581,120,000 \text{ mg}}{24 \text{ hours}} \times \frac{\text{hour}}{60 \text{ minutes}} = \frac{403,556 \text{ milligrams}}{\text{minute}}$$

Answer: 403,556 milligrams/min = 0.89 pound/min.

78. A gas mixture is 80% methane, 15% ethane, 4% propane, and 1% butane (by volume). Their respective LELs and UELs are 5, 3.1, 2.1, 1.86, and 15.0, 12.45, 9.5, and 8.41%. What are the LEL and UEL (in air) for the gas mixture?

Calculations require the application of Le Chatelier's law:

$$LEL = \frac{100}{\dfrac{80}{5} + \dfrac{15}{3.1} + \dfrac{4}{2.1} + \dfrac{1}{1.86}} = 4.30\%$$

$$UEL = \frac{100}{\dfrac{80}{15} + \dfrac{15}{12.5} + \dfrac{4}{9.5} + \dfrac{1}{8.41}} = 14.13\%$$

Answers: LEL = 4.30%. UEL = 14.13%.

79. How much benzene must be evaporated inside a 20.3 L Pyrex® bottle to obtain 50 ppm of benzene vapor? The density of benzene = 0.879 g/mL.

C_6H_6 molecular weight = $(6 \times 12) + 6 = 78$

$$\frac{50}{10^6} \times \frac{20.3\,L}{\underset{\text{gram-mole}}{24.45\,L}} \times \frac{78}{\underset{\text{mL}}{0.879\,g}} = 0.00368\,mL$$

Answer: Inject 3.7 microliters of liquid benzene.

80. How much carbon monoxide gas must be added to a 29.8 L glass bottle to obtain a 35 ppm gas mixture? Assume that the air balance is CO-free by passing ambient air containing 2 ppm CO through a Hopcalite® filter.

CO molecular weight = 12 + 16 = 28.

$$\text{density of CO (at NTP)} = \frac{\dfrac{28\,g}{\text{gram-mole}}}{\dfrac{24.45\,L}{\text{gram-mole}}} = \frac{1.145\,g}{L} = \frac{0.001145\,g}{mL}$$

$$\frac{35}{10^6} \times \frac{29.8\,L}{\underset{\text{gram-mole}}{24.45\,L}} \times \frac{28}{\underset{\text{mL}}{0.001145\,g}} = 1.043\,mL\,CO$$

alternative calculation method: $\dfrac{x}{29,800\,mL} = \dfrac{35}{10^6}$ $x = 1.043\,mL\,CO$

Answer: 1.04 mL of 100% carbon monoxide gas (or, e.g., 10.4 mL of 10% CO).

81. 680 tons of coal containing an average of 0.7 ppm mercury are burned daily in an electricity-generating power plant. Assuming 98% volatilization of every mercury compound in the coal, how much mercury is released every hour?

$$\frac{680 \text{ tons}}{\text{day}} \times \frac{2000 \text{ lb}}{\text{ton}} \times \frac{454 \text{ grams}}{\text{pound}} = 6.17 \times 10^8 \text{ grams of Hg/day}$$

$$\frac{6.17 \times 10^8 \text{ grams/day}}{24 \text{ hours/day}} = 2.57 \times 10^7 \text{ grams/hour}$$

$$0.98 \left[\frac{2.57 \times 10^7 \text{ grams}}{\text{hour}} \right] \times 0.7 \times 10^{-6} \text{ppm Hg} = 17.6 \text{ g Hg/hour}$$

Answer: 17.6 g of mercury emitted per hour (presumably as elemental Hg^o vapor and mercury oxides, sulfides, and sulfates).

82. How many kilograms of sulfur dioxide gas (SO_2) are produced when 6-1/2 tons of coal containing 3.4% sulfur are completely burned (stoichiometrically oxidized)?

$$S + O_2 \rightarrow SO_2 \uparrow (2 S + 3 O_2 \rightarrow 2 SO_3 \uparrow)$$
6.5 tons $\times$ 0.034 = 0.221 ton sulfur = 200.5 kg sulfur

$$\frac{200.5 \text{ kg sulfur}}{32 \text{ g sulfur/gram-mole}} \times \frac{1000 \text{ g}}{\text{kg}} = 6266 \text{ moles of sulfur}$$

Therefore, 6266 mols of sulfur dioxide (SO_2) are produced.
molecular weight of SO_2 = 32 + (16 $\times$ 2) = 64
(6266 mols SO_2) (64 g SO_2/mol) = 401,024 g SO_2 = 401 kg SO_2

Answer: 401 kilograms SO_2, e.g., one ton of sulfur $\Rightarrow$ two tons of SO_2 gas.

83. The decay constant for a reactive gas in air is 4.9 $\times$ 10^{-2} molecular dissociations per minute. What is the half-life of this gas in seconds?

Nt = $N_o e^{-kt}$ first order exponential decay kinetics, where:
Nt = number of molecules remaining after time, t,
N_o = original number of molecules, and
k = the molecular dissociation constant.

substitute $N_o/2$ for N_t and T for t:

$$\frac{N_o}{2} = N_o e^{-kT} \quad \text{solve for T: } 0.5 = e^{-kT}$$

taking natural logs of both sides: $ln\ 0.5 = -kT$

thus: $T = \dfrac{ln\ 0.5}{-k} = \dfrac{ln\ 0.5}{-\left(4.9 \times 10^{-2} \text{ dissociations / minute}\right)} = 14.15 \text{ minutes}$

Answer: 849 sec, i.e., 50% decay every 849 sec.

84. What fraction of the gas in problem 83 remains after one hr?

$N_t = N_o e^{-kt}$ dividing both sides by N_o:

$$\frac{N_t}{N_o} = e^{-kt} \qquad \frac{N_t}{N_o} = e^{-(0.049/\text{minute})(60 \text{ minutes})} = 5.286 \times 10^{-2} = 5.3\%$$

Answer: 5.3% of the unstable gas remains in the air after one hr.

85. What are the mean and the standard deviation for the following analytical results of the amounts of beryllium (in micrograms) on 10 cm × 10 cm surface wipe samples:

6.3, 9.7, 9.4, 12.1, 8.5, and 7.7?

Answers: Mean (average) = 9.0 mcg Be/100 cm². Standard deviation = 1.8.

86. In question 85, what range most likely covers about 95% of the results?

Answer: 5.4 – 12.6 mcg Be/100 cm².

87. What is the statistical correlation between the following corresponding "x" and "y" values: x = 5, 13, 8, 10, 15, 20, 4, 16, 18, and 6; y = 10, 30, 30, 40, 60, 50, 20, 60, 50, and 20?

Answer: r = 0.866.

	x	**y**
x	1	
y	**0.86587**	1

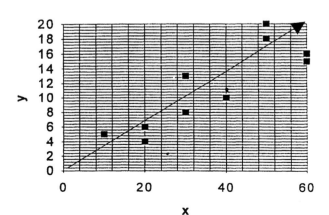

88. Workers inhale respirable dust which is 27% quartz and 11% cristobalite. What is the permissible exposure limit for this fibrogenic dust mixture?

$$PEL = \frac{10 \text{ mg/M}^3}{2 + \% \text{ quartz} + 2\,(\% \text{ cristobalite})} = \frac{10 \text{ mg/M}^3}{2 + 27 + (2 \times 11)} = 0.196 \text{ mg / M}^3$$

Answer: < 0.2 mg/M^3, assuming the balance of the dust mixture is biologically inert.

89. A 2 mL CS$_2$ extract of a small charcoal tube contained 3.7 mcg EO per mL. Air was sampled at 93 mL/min for 97 min at 77°F. What was the ethylene oxide concentration in air assuming 79% analytical recovery? The air sample was taken in Boston.

EO = C$_2$H$_4$O   molecular weight = 24 + 4 + 16 = 44

(93 mL/min) × 97 min = 9021 mL = 9.021 L

2 mL × (3.7 mcg/mL) = 7.4 mcg EO

$$\frac{\dfrac{7.4 \text{ mcg EO}}{9.021 \text{ L}} \times 24.45}{44} = 0.45 \text{ ppm} \qquad \frac{100}{79} = 1.266 \quad 0.45 \text{ ppm} \times 1.266 = 0.6 \text{ ppm}$$

Answer: 0.6 ppm ethylene oxide gas.

90. Determine the cubic feet of vapor produced from the evaporation of one gallon of methanol at 60°F in New Orleans. The density of methyl alcohol = 0.8 g/mL.

one gallon = 3785 mL × 0.8 g/mL = 3028 g = 6.67 pounds

CH$_3$OH molecular weight = 12 + 16 + 4 = 32.

1 pound CH$_3$OH forms (359 ft^3/32) vapor at 0°C = 11.2 ft^3

at 60°F: °R = °F + 460 = 520°R

°R = 32°F + 460 = 492°R

$$\frac{\text{ft}^3}{\text{lb}} = \frac{(520°R)\,(359 \text{ ft}^3)}{(492°R)\,(\text{mol. wt.})} = \frac{379.4}{32} = \frac{11.86 \text{ ft}^3}{\text{lb}} \qquad 6.67 \text{ lb} \times \frac{11.86 \text{ ft}^3}{\text{lb}} = 79.1 \text{ ft}^3$$

Answer: 79 ft^3 of vapor are produced by evaporating one gallon of CH$_3$OH.

91. A four-in diameter Petri dish had 0.6 colony-forming units/cm^2 after culture incubation from a 27.4 ft^3 air sample. How many viable organisms/M^3 does this represent?

27.4 ft^3 × 28.3 L/ft^3 = 775.4 L = 0.775 M^3

4-in diameter = 2-in radius, r, = 5.08 cm

$A = \pi r^2 = \pi (5.08 \text{ cm})^2 = 81.07 \text{ cm}^2$

$81.07 \text{ cm}^2 \times 0.6 \text{ CFU/cm}^2 = 48.6 \text{ CFU/Petri dish}$

$$\frac{48.6 \text{ CFU}}{0.775 \text{ M}^3} = \frac{63 \text{ CFU}}{\text{M}^3}$$

Answer: 63 colony-forming units per cubic meter of air.

92. 0.062 mL of ethyl acetate is evaporated/min into a 34 L/min air stream. What is the EA vapor concentration in ppm at 77°F and 760 mm Hg? Molecular weight of EA = 88. Density of liquid EA = 0.9 g/mL.

$0.062 \text{ mL} \times (0.9 \text{ g/mL}) = 0.0558 \text{ gram}$ $\dfrac{55.8 \text{ mg}}{34 \text{ L}} = \dfrac{1.64 \text{ mg}}{\text{L}} = \dfrac{1640 \text{ mg}}{\text{M}^3}$

$$\text{ppm} = \frac{\dfrac{1640 \text{ mg}}{\text{M}^3} \times 24.45}{88} = 456 \text{ ppm ethyl acetate}$$

Answer: 456 ppm ethyl acetate vapor.

93. A direct-reading instrument to measure CO gas in air is calibrated at sea level and 25°C. What is the corrected reading for the instrument if it indicates 47 ppm CO at 7500 ft and 29°C?

"Rule of thumb": For every 1000 ft of elevation above sea level, the barometric pressure decreases by about one in of mercury. Therefore, for 7500 ft, the equivalent pressure = 7.5-in. Hg × (25.4 mm/in) = 190.5 mm Hg.

760 mm Hg – 190.5 mm Hg = 569.5 mm Hg.

$$\text{ppm at NTP} = \text{meter reading} \times \frac{P}{760} \times \frac{298 \text{ K}}{T} =$$

$$47 \text{ ppm} \times \frac{P}{760} \times \frac{298 \text{ K}}{T} = 47 \text{ ppm} \times \frac{569.5 \text{ mm Hg}}{760 \text{ mm Hg}} \times \frac{298 \text{ K}}{302 \text{ K}} = 34.8 \text{ ppm}$$

Answer: 35 ppm. Instrument correction factor = $\dfrac{34.8 \text{ ppm CO}}{47 \text{ ppm CO}} = 0.74.$

94. The concentration of a ketone vapor in inhaled air is 48 ppm. It is 19 ppm in the end-exhaled air of a worker. What is the average body retention of this ketone?

$$\% \text{ retention} = \frac{C_i - C_e}{C_i} \times 100 = \frac{48 \text{ ppm} - 19 \text{ ppm}}{48 \text{ ppm}} \times 100 = 60\%$$

Answer: 60%. "Retention" is a function of metabolism-excretion, detoxification, and storage.

95. Two airborne dust samples obtained in a worker's breathing zone had the following results for combined daily exposure to respirable silica:

air sample	duration, minutes	liters	respirable weight, mg	mg/M³	silica
morning	161	274	0.961	3.51	6.9% quartz
					1.8% cristobalite
					0.0% tridymite
afternoon	247	420	0.530	1.26	7.3% quartz
					1.9% cristobalite
					0.0% tridymite
total	408	694	1.491		

Calculate the percent quartz, cristobalite, and tridymite in the respirable particulate fractions. Calculate the PEL for the mixture and the employee's exposure. Adjust his exposure to an eight-hr time-weighted average. Assume that the remainder of the worker's exposure was in an area with no airborne silica dust. Also assume that the clean-shaven worker used an organic vapor cartridge respirator.

$$\text{quartz: } 6.9\% \times \frac{0.961\,\text{mg}}{1.491\,\text{mg}} + 7.3\% \times \frac{0.530\,\text{mg}}{1.491\,\text{mg}} = 7.0\%$$

$$\text{cristobalite: } 1.8\% \times \frac{0.961\,\text{mg}}{1.491\,\text{mg}} + 1.9\% \times \frac{0.961\,\text{mg}}{1.491\,\text{mg}} = 2.4\%$$

$$\text{PEL} = \frac{10\,\text{mg/M}^3}{2 + \%\,\text{quartz} + 2(\%\,\text{cristobalite}) + 2(\%\,\text{tridymite})} =$$

$$\frac{10\,\text{mg/M}^3}{2 + 7 + 2(2.4) + 2(0)} = \frac{0.72\,\text{mg}}{\text{M}^3}$$

$$\text{actual exposure} = \frac{(0.961\,\text{mg/M}^3) + (0.530\,\text{mg/M}^3)}{0.694\,\text{M}^3} = 2.15\,\text{mg/M}^3$$

$$\text{TWAE} = \frac{2.15\,\text{mg}}{\text{M}^3} \times \frac{408\,\text{minutes}}{480\,\text{minutes}} = 1.83\,\text{mg/M}^3$$

Answers: Quartz = 7.0%. Cristobalite = 2.4%. PEL = 0.72 mg/M³. Exposure period = 2.15 mg/M³. Adjusted 8-hr TWAE = 1.83 mg/M³.

96. A precision rotameter was calibrated at 1.7 L/m at sea level and 70°F. Air will be sampled for respirable silica dust at 90°F and at a barometric pressure of 633 mm Hg with a cyclone dust sampler. What corrected rotameter reading must be used with this cyclone? What is the rotameter correction factor?

$$90°F = 32.2°C \quad 32.2°C + 273 \text{ K} = 305.2 \text{ K}$$

$$\frac{1.7\,L}{min} \times \frac{633 \text{ mm Hg}}{760 \text{ mm Hg}} \times \frac{298 \text{ K}}{305.2 \text{ K}} = \frac{1.33\,L}{min}$$

Answers: 1.38 L/min. $C_f = \dfrac{1.38\,L/m}{1.7\,L/m} = 0.812$.

97. A clothing dry cleaning plant purchases two 55-gallon drums of perchloro-
    ethylene every month. Losses to the environment are entirely due to evapora-
    tion. If 75% of the total loss is through a vent exhausting 200 cfm, what is
    the average "perc" vapor concentration in the vent if the plant operates six
    days/week and nine hrs daily. The density of liquid "perc" = 1.62 g/mL.

    110 gallons/month × 3.785 L/gallon = 416.4 L "perc" evaporated every month

$$\frac{6 \text{ days}}{week} \times \frac{9 \text{ hours}}{day} \times \frac{60 \text{ minutes}}{hour} \times \frac{4.5 \text{ weeks}}{month} \times \frac{200 \text{ ft}^3}{minute} = \frac{2.916 \times 10^6 \text{ ft}^3}{month}$$

$$\frac{416.4 \text{ L "perc"/month}}{2,916,000 \text{ ft}^3 / \text{month}} = \frac{416,400 \text{ mL "perc"}}{82,581,120 \text{ L of air and vapor}} = \frac{0.05 \text{ mL}}{L} = \frac{5 \text{ mL}}{M^3}$$

    5 mL × (1.62 g/mL) = 8.1 g of "perc"   8.1 g/M³ = 8100 mg/M³

    molecular weight of "perc" = (4 × 35.5) + (2 × 12) = 166

$$ppm = \frac{\dfrac{8100 \text{ mg}}{M^3} \times 24.45}{166} = 1193 \text{ ppm} \quad 1193 \text{ ppm "perc"} \times 0.75 = 895 \text{ ppm}$$

    Answer: ≅ 900 ppm "perc" vapor.

98. How much water vapor is in every 1000 ft³ of building air if the relative
    humidity is 50% at an air temperature of 80°F? The vapor pressure of H_2O at
    80°F = 26.2 mm Hg. Assume that this building is located in Norfolk, VA.

$$80°F = 26.67°C \quad 26.67°C + 273 \text{ K} = 299.76 \text{ K}$$

$$\frac{22.4 \text{ L}}{gram\text{-}mole} \times \frac{299.76 \text{ K}}{273 \text{ K}} = \frac{24.59 \text{ L}}{gram\text{-}mole}$$

$$ppm\ H_2O \text{ vapor} = 0.5 \times \frac{26.2 \text{ mm Hg}}{760 \text{ mm Hg}} \times 10^6 = 17,237 \text{ ppm}$$

$$\frac{mg}{M^3} = \frac{ppm \times \text{mol. wt.}}{24.59 \text{ L/gram-mole}} = \frac{17,237 \times 18}{24.59} = 12,612 \text{ mg/M}^3$$

$$\frac{12.61 \text{ grams}}{35.3 \text{ ft}^3} = \frac{0.357 \text{ gram}}{\text{ft}^3}$$

$$\frac{357 \text{ g}}{10^3 \text{ ft}^3} = \frac{0.79 \text{ lb}}{10^3 \text{ ft}^3}$$

Answer: 0.79 pound of water vapor per 1000 ft^3 of air.

99. Determine the amount (in grams) of CO_2 produced by completely burning 25 L of ethane in air at STP according to the stoichiometric combustion equation:

$$2 \text{ C}_2\text{H}_6 + 7 \text{ O}_2 \rightarrow 4 \text{ CO}_2 \uparrow + 6 \text{ H}_2\text{O} \uparrow$$

C_2H_6 molecular weight = $(2 \times 12) + 6 = 30$

30 g/g-mol   22.4 L = 30 g   25 L = $\times$ g

$$\frac{22.4 \text{ L}}{25 \text{ L}} = \frac{30 \text{ grams}}{\text{x grams}} \qquad \frac{33.48 \text{ g}}{30 \text{ g / g-mole}} = 1.116 \text{ moles of C}_2\text{H}_6 \qquad \frac{2}{7} = \frac{1}{3.5}$$

1.116 mols $\times$ 2 = 2.232 mols of CO_2 produced

2.232 mols of CO_2 $\times$ (44 g/g-mol) = 98.2 g

Answer: 98.2 g of CO_2 are produced.

100. If the concentration of a hydrocarbon vapor in inhaled air is 70 mg/M^3, exposure is eight hrs, respiratory ventilation rate is 1.2 M^3/hr, and the average retention is 73%, what is the biologically-absorbed dose?

C (mg/M^3) $\times$ T (hrs) $\times$ V (M^3/hr) $\times$ R (%) = absorbed dose in milligrams =

$$\frac{70 \text{ mg}}{\text{M}^3} \times 8 \text{ hours} \times \frac{1.2 \text{ M}^3}{\text{hour}} \times 0.73 = 490 \text{ milligrams}$$

Answer: 490 milligrams.

101. Workers in chemical plants such as petroleum refineries often work in six week schedules of three 12-hr days for three weeks followed by four 12-hr work days for three weeks. What is the PEL and TLV reduction for such workers?

Note that the weekly average exposures of 36 and 48 hrs are slightly different than a normal work week schedule of 40 hrs.

$$\text{TLV reduction factor} = \frac{8 \text{ hours}}{12 \text{ hours}} \times \frac{24 \text{ hours} - 12 \text{ hours}}{16 \text{ hours}} = 0.5$$

Answer: 0.5. Note that the reduction factor applies to 12-hr work days regardless if the exposures are three, four, or five or more days per week.

102. Hydrogen chloride has a TLV (C) of five ppm. What is the modified TLV for HCl gas for a work schedule of 12 hrs per day for a three-day work week?

    Answer: Since the basis of the TLV is prevention of acute respiratory irritation, lowering of the limit might not be justifiable. The concept of reducing eight-hr limits generally applies to systemically-toxic agents and not normally to those air contaminants having acute effects and with C, or ceiling, limits.

103. Determine the mass of vapor expelled when 1000 gallons of benzene are added to a 5000 gallon storage tank if the temperatures of the tank space and benzene are 20°C. Assume that the vessel wall is warmer than the liquid benzene, the tank is in Baltimore, and the tank contained 2500 gallons of benzene for the past three weeks. The vapor pressure of benzene at 20°C = 75 mm Hg.

    The American Petroleum Institute provides a formula addressing the release of solvent vapors from tanks during filling. This is used to estimate vapor emissions into work room air by displacement of saturated vapors of volatile hydrocarbons.

    The mass of vapor expelled by displacement when a volatile liquid is transferred into a tank using the API formula is:

$$M = 1.37 \, VSP_v \left[\frac{MW}{T} \right],$$

    where:

    M = mass of vapor expelled, lb
    V = volume of liquid transferred to tank, ft³
    S = fraction of vapor saturation of expelled air
    P_v = vapor pressure of liquid, atmospheres
    MW = molecular weight of solvent or hydrocarbon
    T = temperature of tank vapor space, °R

    For splash filling of a tank initially free of vapors or for the refilling of tanks from which the same liquid was just withdrawn, the value of S is normally 1.0.

    V = 1000 gallons = 3785 L = 133.7 ft³
    P_v = 75 mm Hg   75 mm Hg/760 mm Hg = 0.09868 atmosphere
    C_6H_6 = molecular weight = (6 × 12) + 6 = 78   20°C = 68°F
    °R   = °F + 459.69 = 68°F + 459.69 = 527.69°F

$$M = (1.37)\left(133.7 \text{ ft}^3\right)(1.0)(0.09868 \text{ atm})\left[\frac{78}{527.69}\right] = 2.67 \text{ lb}$$

    alternate calculation method:

1000 gallons benzene vapor = 3785 L of benzene vapor

$$\text{ppm} = \frac{75 \text{ mm Hg}}{760 \text{ mm Hg}} \times 10^6 = 98,684 \text{ ppm}$$

$$\frac{\text{mg}}{\text{M}^3} = \frac{\text{ppm} \times \text{mol. wt.}}{24.04} = \frac{98,684 \text{ ppm} \times 78}{24.04} = \frac{98,684 \times 78}{24.04} = 320,190 \text{ mg/M}^3$$

$$(320.19 \text{ g/M}^3) \times 3.785 \text{ M}^3 = 1211.9 \text{ g} = 2.67 \text{ lb}$$

Answer: 2.67 pounds of benzene vapor are expelled to the atmosphere.

104. Workers are exposed to systemically-toxic air contaminants 12 hrs/day and 6 days/week. By how much should the exposure limits be reduced to help protect their health?

$$\frac{40}{h} \times \frac{168 - h}{128},$$

where:

h = hours worked/week
40 = hours in a normal work week
168 = total hours in a week
128 = hours available in a normal work week to detoxify, excrete toxicants, etc.

$$\frac{40}{72} \times \frac{168 - 72}{128} = 0.42$$

Answer: Reduce their exposure limits (including action levels) by at least 42%.

105. Using the OSHA model, what is the equivalent PEL for cumulative poisons such as lead and mercury (PELs = 0.05 mg/M^3) if exposures are, for example, 50 hrs per week?

$$\text{equivalent PEL} = \frac{40 \text{ hours}}{\text{exposure hours/week}} = \text{weekly adjustment}$$

e.g., $\frac{40 \text{ hours}}{50 \text{ hours}} \times 0.05 \text{ mg/M}^3 = 0.04 \text{ mg/M}^3$

Answer: 0.04 mg Pb or Hg, e.g.,/M^3.

106. How many molecules of mercury vapor exist per cm^3 in cold traps of mercury diffusion pumps maintained at minus 120°C? The vapor pressure of mercury at this temperature = 10^{-16} mm Hg.

$$\frac{\text{moles of Hg}}{\text{cm}^3} = n = \frac{PV}{RT} = \frac{10^{-16}\ \text{atmosphere} \times 10^{-3}\text{L}}{0.082\ \text{L-atm}^{-1}\ \text{mole}^{-1} \times 153°\text{K}} = 1 \times 10^{-23}\ \text{mole Hg}$$

10^{-23} mol Hg $\times$ 6.023 $\times$ 10^{23} molecules/mol = 6 molecules of Hg/cm^3

Answer: Six mercury molecules per cubic centimeter. Contrast this to the 1.5 $\times$ 10^{11} mercury molecules in each cm^3 of air at the PEL of 50 mcg Hg/M^3.

107. Determine the weight of solvent vapor emitted per day from a paint baking oven drying 40 gallons/day as a blend of four gallons of unthinned enamel to one gallon of thinner. The enamel weighs 9.2 lb/gallon of which 51% is volatile. The thinner weighs seven lb/gallon. Assume no particulates form and all volatiles exist in the vapor phase.

This is a blend of four gallons unthinned enamel plus one gallon of thinner. Thus, (40/5) $\times$ 4 = 32 gallons of unthinned enamel are used daily plus (40/5) $\times$ 1 = 8 gallons of thinner are used daily.

volatiles in the unthinned enamel are:

$$32\ \text{gallons} \times \frac{9.2\ \text{lb}}{\text{gallon}} \times (1 - 0.49) = 150\ \text{lb}$$

volatiles in the thinner are: (8 gallons) (7 lb/gallon) = 56 lb

150 lb + 56 lb = 206 lb/day

Answer: 206 pounds of solvent vapor are emitted to the atmosphere daily.

108. The solvent vapor concentration in a plant is 130 ppm and cannot be further reduced by improved work practices. It is necessary to reduce this to at least 15 ppm to protect the health of the exposed workers. By how much must the dilution air be increased?

If the solvent vapor contaminant is generated at a steady rate:

$$\frac{Q}{Q_o} = \frac{C}{C_o} \quad \frac{C_o}{C} = \frac{15\ \text{ppm}}{130\ \text{ppm}}, \text{therefore:} \quad \frac{Q}{Q_o} = 8.7, \text{i.e.,}$$

the supplied air volume must be increased at least 8.7 times.

Answer: At least 8.7 times more clean dilution air with excellent air mixing.

109. A 100 $\times$ 50 $\times$ 12-ft work shop is provided with two cfm of outside air per ft^2 of floor area. A gaseous contaminant is evolved in this shop at a rate of 6.6 cfm. An air sample taken when the exhaust fan is shut off reveals a concentration of 580 ppm. How long must this fan operate before the air concentration

reaches 100 ppm? Assume excellent mixing of the dilution air with the contaminated air.

$100 \times 50 \text{ ft} \times (2 \text{ cfm/ft}^2) = 10,000 \text{ cfm}$

$$t, \text{ minutes} = \frac{2.303 \times \text{ft}^3}{Q} \times \log \frac{G - (Q \times C_a)}{G - (Q \times C_b)}$$

$$t = \frac{2.303 \times 60,000 \text{ ft}^3}{10,000 \text{ cfm}} \times \log \frac{6.6 \text{ cfm} - \left(10,000 \text{ cfm} \times \frac{100}{10^6}\right)}{6.6 \text{ cfm} - \left(10,000 \text{ cfm} \times \frac{580}{10^6}\right)}$$

$$13.82 \times \log \left[\frac{6.6 - 1}{6.6 - 5.8}\right] = 13.82 \times \log 7 = 13.82 \times 0.845 = 11.7 \text{ minutes}$$

Answer: At least 12 min if there is good mixing — longer if there is poor mixing of fresh air with pockets of contaminated air.

110. Ambient air has been reported to contain 0.01 to 0.02 microgram of mercury per cubic meter. How many molecules of mercury are in each liter of air inhaled at this concentration?

$$\frac{6.023 \times 10^{23} \text{ molecules Hg/gram-mole}}{200.59 \text{ g Hg/gram-mole}} = \frac{3 \times 10^{21} \text{ molecules}}{\text{gram}}$$

$$= \frac{3 \times 10^{15} \text{ molecules}}{\text{microgram}}$$

$0.01 \text{ mcg/M}^3 = 0.00001 \text{ mcg/L}$

$$\frac{10^{-5} \text{ mcg}}{L} \times \frac{3 \times 10^{15} \text{ molecules}}{\text{mcg}} = \frac{3 \times 10^{10} \text{ molecules Hg}}{L}$$

Answer: 3×10^{10} molecules of Hg per liter of air.

111. Seven mL of 100% carbon monoxide gas were added by a gas syringe to a plastic bag into which 127 L of CO-free air had been metered. What is the resultant CO gas concentration?

$$\text{ppm CO} = 10^6 \times \frac{0.007 \text{ L}}{127 \text{ L} + 0.007 \text{ L}} = 55.1 \text{ parts CO}/10^6 \text{ parts of air}$$

$$\text{alternative calculation method:} \quad \frac{7 \text{ mL}}{127,007 \text{ mL}} \times \frac{x}{10^6} \qquad x = 55 \text{ ppm}$$

Answer: 55 ppm CO.

112. An incinerator produces 0.032 lb of phosgene gas per hour for every 100 lb of pentachlorophenol processed per hour at extreme temperatures with methane and excess air. If the total gas flow rate leaving the incinerator is 4600 cfm at NTP, what is the average phosgene gas concentration in the exhaust gas?

$C_6OHCl_5 + CH_4$ + excess air $(O_2) \rightarrow CO_2$, CO, Cl_2, HCl, H_2O, etc. + $COCl_2$

4600 cfm = 276,000 ft³/hr = 7.81×10^6 L/hr

$$\frac{0.032 \text{ lb/hour}}{276,000 \text{ cfh}} = \frac{14.53 \times 10^6 \text{ mcg}}{7.81 \times 10^6 \text{ L}} = 1.86 \text{ mcg/L}$$

molecular weight $COCl_2$ = 12 + 16 + 2 (31.5) = 99

$$ppm = \frac{\dfrac{1.89 \text{ mcg}}{L} \times 24.45}{99} = 0.46 \text{ ppm} = 460 \text{ ppb}$$

Answer: 0.46 ppm $COCl_2$ gas.

113. A worker was exposed to a 6 mg/M³ dust cloud of radioactive mercuric cyanide [Hg^{203} $(CN)_2$] for 30 min. Assuming no other exposures that day, what were his TWAEs to mercury and cyanide? Hg^{203} is a β-emitter with a $T_{1/2}$ of 47 days. What are the three health hazards? What is the greatest health hazard? What is the least acute? Assume that the dust is entirely respirable with all particles less than three microns.

molecular weight $Hg(CN)_2$ = 203 + (12 × 2) + (14 × 2) = 255
(203/255) × 100   = 79.6% Hg   (100 – 79.6%) = 20.4% CN⁻
(6 mg/M³) × 0.796   = 4.78 mg Hg^{203}/M³
(6 mg/M³) × 0.204   = 1.22 mg CN⁻/M³

$$\frac{\left(4.78 \text{ mg Hg/M}^3\right) \times 0.5 \text{ hour} + 0}{8 \text{ hours}} = 0.30 \text{ mg Hg/M}^3$$

(TLV inorganic Hg = 0.05 mg Hg/M³)

$$\frac{\left(1.22 \text{ mg CN}^{-2}/M^3\right) + 0}{8 \text{ hours}} = 0.08 \text{ mg CH}^{-2}/M^3 \quad \left(\text{TLV CN}^{-2} = 5 \text{ mg/M}^3\right)$$

Answer: 0.30 mg Hg/M³. 0.08 mg CN⁻/M³. Acute inorganic mercury and cyanide poisoning and internal ionizing radiation (pulmonary deposition and systemic absorption of an internal β-emitter). The ionizing radiation hazard more than likely outweighs the mercury and cyanide hazards because of delayed, chronic effects (Ca?). Although the mercury exposure exceeded the

TLV by six times, it was a single occurrence, and the long term, chronic effects from this are probably negligible.

114. 13.7 L of air at 19°C and 741 mm Hg were sampled in a midget impinger containing 14.4 mL of collection solution with 100% absorption efficiency for SO_2. The SO_2 concentration was analyzed at 11.6 micrograms/mL. Calculate the ppm of SO_2 in the air sample.

total SO_2 sampled = (11.6 mcg/mL) × 14.4 mL = 167 mcg

the volume of 1 micromol SO_2 gas at 19°C and 741 mm Hg =

$$1.0 \text{ micromole } SO_2 = \frac{224 \text{ microliters}}{\text{micromole}} \times \frac{760 \text{ mm}}{741 \text{ mm}} \times \frac{292°K}{273°K} = 24.6 \text{ microliters } SO_2$$

$$SO_2 \text{ concentration in ppm} = \frac{167 \text{ mcg } SO_2}{13.7 \text{ L air}} \times \frac{\text{micromole } SO_2}{64 \text{ mcg } SO_2} \times$$

$$\frac{24.57 \text{ microliters } SO_2}{\text{micromole } SO_2} = \frac{4.68 \text{ microliters } SO_2}{\text{liter of air}}$$

Answer: 4.7 ppm SO_2.

115. One drop (0.05 mL) of TDI evaporates in an unventilated, closed telephone booth. What is the concentration of TDI vapor if the interior volume of the booth is 1.5 M^3? Is this atmosphere hazardous to health? TDI molecular weight = 174. The density of liquid TDI = 1.22 g/mL.

0.05 mL × 1.22g/mL = 0.061 g = 61 milligrams of TDI

$$ppm = \frac{\frac{mg}{M^3} \times 24.45}{\text{molecular weight}} = \frac{\frac{61 \text{ mg}}{1.5 \text{ M}^3} \times 24.45}{174} = 5.7 \text{ ppm} = 5700 \text{ ppb}$$

(5700 ppb/5 ppb PEL) = 1140 times > PEL. LC_{Lo} = 500 ppb.

Answers: 5.7 ppm, or 1140 times the PEL and 11.5 times the LC_{Lo}.

116. How long must one sample at two L/min with a 25 mm diameter cellulose ester membrane filter to evaluate asbestos fibers at 0.04 fiber per cubic centimeter of air? The filter area is 385 mm^2. A minimum fiber count is taken as 100 per mm^2.

$$\text{sampling time (minutes)} = \frac{385 \text{ mm}^2 \times \frac{100 \text{ fibers}}{\text{mm}^2}}{\frac{2.0 \text{ L}}{\text{minute}} \times \frac{0.04 \text{ fiber}}{\text{cc}} \times \frac{1000 \text{ cc}}{\text{L}}} = 481.25 \text{ minutes}$$

Answer: 481 min (= eight hrs), i.e., this is a full work shift breathing zone air sample to determine asbestos exposure at 20% or greater of the OSHA PEL (0.04 f/cc).

117. Air contaminants in a 50,000 gallon tank are 80 ppm CO, three ppm HCN, and eight ppm H_2S. Three pipe fitters have work to do in the tank. Would you:

    a. Permit entry to not exceed 30 min every eight hrs?
    b. Permit entry to not exceed 60 min every eight hrs?
    c. Permit entry to not exceed more than 25% of the work shift?
    d. Permit entry only if the pipe fitters are outside for one hr for every hour that they are inside the tank?
    e. Ventilate the tank with a 500 cfm blower for 20 min?
    f. Permit entry only if the pipe fitters have escape masks?
    g. Not permit entry?
    h. Duck and cover?
    i. Contact a board-certified industrial hygienist?

    Answer: g. First, it would be highly unusual to have three chemical asphyxiant gases exist in the same confined space. Second, since it is likely that the pipe fitters could breach the integrity of the pipes and tanks containing these gases, much higher gas concentrations might occur in their work area. Because of this, entry should not be permitted until further evaluation is done. All elements of a comprehensive confined space entry procedure are required to protect the health and safety of these pipe fitters (refer to OSHA 29 *CFR* 1910.1025).

118. Determine the volume and mass conversion of ozone gas at 25°C and 760 mm Hg.

    molecular weight of $O_3 = 3 \times 16 = 48$

    $T = 273\ K + 25°C = 298\ K \qquad R = 0.0821$ L-atm/mol-K

    $$ppm = \frac{W}{V} \times \frac{RT}{PM} \times 10^6 = \frac{W(g) \times 10^6 (mcg/g)}{V(L) \times 10^{-3} M^3/L}$$

    $$\frac{(0.0821\ L\text{-atm}/mole\text{-}K) \times 298\ K}{1\ atmosphere \times (48\ g/mole) \times 10^6} \qquad 1\ ppm = 1960\ mcg/M^3$$

    $1\ mcg/M^3 = 0.51 \times 10^{-3}\ ppm = 0.51\ ppb$

    Answer: 1 ppm = 1960 mcg/M^3. 1 $mcg/M^3 = 5.1 \times 10^{-4}$ ppm.

119. Calculate the dust emission from an incinerator stack with the following stack sampling conditions:

V_m	= volume of gas by meter	= 105 ft^3
T_m	= temperature at meter	= 83°F + 460 = 543°A
P_b	= barometric pressure	= 27.8-in Hg
P_m	= average suction at meter	= 2.5-in Hg
V_w	= condensed water	= 138 cc
W_t	= filtered dust	= 23 g
V_o	= Pitot traverse exhaust volume	= 37,200 cfm

volume of total gas samples (at meter conditions): convert H_2O condensate to water vapor volume at meter conditions:

$$V_v = 0.00267 \times \frac{V_w \times T_m}{P_b \times P_m} = 0.00267 \times \frac{138\ cm^3 \times 543°\ A}{24.8'' - 2.5''} = 7.92\ ft^3$$

total gas sampled = $V_v + V_m$ = 7.92 ft^3 + 105 ft^3 = 112.92 ft^3

moisture content of the gas sampled (by volume) = M_m = volume of the moisture remaining in the metered gas at meter conditions in ft^3

vapor pressure of H_2O at 83°F = 1.138-in Hg

$$M_m = \frac{V_p \times V_m}{P_b - P_m} = \frac{1.138''\ Hg \times 105\ ft^3}{27.8''\ Hg - 2.5''\ Hg} = 4.72\ ft^3$$

$$\%\ moisture = \frac{V_v + M_m}{V_v + V_m} \times 100 = \frac{7.92\ ft^3 + 4.72\ ft^3}{7.92\ ft^3 + 105\ ft^3} \times 100 = 11.2\%$$

Convert total sampled gas volume to stack conditions:

$$V_t = (V_m + V_v) \times \frac{P_b - P_m}{P_s} \times \frac{T_s}{T_m} = (105\ ft^3 + 7.92\ ft^3) \times \frac{27.8''\ Hg - 2.5''\ Hg}{27.8''\ Hg} \times$$

$$\frac{700°\ A}{543°\ A} = 132.4\ ft^3$$

Dust concentration is generally referred to as grains per cubic foot and pounds per hour. To express concentration in grains/ft^3, divide filtered dust weight (W_t) by total sampled gas at stack conditions (V_t), and convert grams to grains:

grains/ft^3 = (W_t/V_t) × 15.43 = (23/132.4) × 15.43 = 2.68 grains/ft^3

lb/hr = (grains/ft^3) × V_o × (60/70) =

$$\frac{2.68\ grains}{ft^3} \times 37,200\ cfm \times \frac{60}{760} = \frac{854\ lb}{hr}$$

Answer: 854 pounds of dust are emitted from the stack per hour.

120. 6.1 mL of trichloroethylene are evaporated in a sealed container with dimensions of (1.65 × 1.65 × 2.35) meters. Chamber air and wall temperature are 25.5°C. The barometric reading is 730 mm Hg. The density of TCE is 1.46 g/mL. What is the TCE vapor concentration in the container?

molecular weight of $CHCl = CCl_2 = (2 \times 12) + 1 + (3 \times 35.5) = 131.5$

6.1 mL × 1.46 g/mL = 8.906 g of liquid TCE

$$\text{molar volume} = 24.45\,L \times \frac{760\,\text{mm Hg}}{730\,\text{mm Hg}} \times \frac{298.5\,K}{298\,K} = 25.5\,\text{liters}$$

$$ppm = \frac{10^6 \times \dfrac{\dfrac{8.906\,g}{131.5\,\text{grams}}}{\text{gram-mole}}}{1000 \times \dfrac{(1.65 \times 1.65 \times 2.35)\,\text{meters}}{25.5\,L}} = 269.9$$

Answer: 270 ppm TCE vapor.

121. A drying process using acetone is done on an open bench in the center of a 20 × 20 × 10–ft room. The room has two to three air changes every hour. Between seven and 10 gallons of acetone evaporate every eight hours as determined by the department's solvent purchase records. What are the hazards? Risks? The vapor volume of acetone is 44 ft³ per gallon. Calculate the average concentration of acetone vapor in the room assuming good mixing of outside air with the vapor contaminated air.

Estimate the hazards and the risks by assuming the worst case situation, i.e., the evaporation of 10 gallons every eight hrs and an air exchange rate of two room volumes every hour.

20 × 20 × 10 ft = 4000 ft³

$$4000\,ft^3 \times \frac{2\,\text{air changes}}{\text{hour}} \times 8\,\text{hours} = \frac{64,000\,ft^3}{8\,\text{hours}}$$

total 100% acetone vapor volume = (44 ft³/gallon) × 10 gallons = 440 ft³

total volume of air required to dilute to the TLV and PEL =

$$\frac{\text{volume of 100\% solvent vapor} \times 10^6}{\text{TLV (in ppm)}} = \frac{440\,ft^3 \times 10^6}{750\,ppm} = 58,670\,ft^3$$

$$\frac{58,760\,ft^3 \times 750\,ppm}{64,000\,ft^3} = 688\,ppm$$

Answer: About 700 ppm acetone vapor. Use local exhaust ventilation. *Note:* Use the lowest air exchange rate and the highest acetone consumption rate in the calculations. Apply a generous safety factor, e.g., 10, to ensure that the vapor levels are substantially below the action level and as low as technically and economically feasible. Give careful consideration to improved work practices, alternative drying techniques, using mechanical local exhaust ventilation, the use of respirators, etc.

122. The major component of natural gas is methane (CH_4) which burns in excess air (i.e., excess oxygen) according to the reaction:

$$CH_4 + 2\ O_2 + 2\ (3.78)\ N_2 \rightarrow CO_2\uparrow + 2\ H_2O\uparrow + 7.56\ N_2\uparrow$$

which is: one mol of methane + 9.56 "moles" of air yields one mol of CO_2 + two mols of H_2O + 7.56 mols of N_2.

Determine the weight and volume of air required to burn 1000 ft³ of CH_4 if the gas and air are at 70°F and 15 pounds per square in absolute pressure.

According to the reaction, 9.56 "mols" of air are required for each mole of fuel gas. Since the molar volume of both gases is the same at the same temperature and pressure, (9.56) (1000 ft³) = 9560 ft³ of air are needed.

The molar volume at 70°F and 15 psia is:

$$358\ ft^3 \times \frac{70°F + 460°F}{32°F + 460°F} \times \frac{14.7\ lb/in^2}{15\ lb/in^2} = 378\ ft^3$$

The molar volume is 358 ft³ at 32°F and 14.7 lb per square in absolute, i.e., this is the volume of a pound-mol of any ideal gas or vapor at these conditions.

$$\text{Therefore, the weight of air required} = \frac{9560\ ft^3}{378\ ft^3/pound\text{-}mole}$$

$$= 25.3\ \text{"moles" air.}$$

29 = the "apparent" molecular weight of air

(25.3 mols) (29 lbs/"mol" of air) = 734 pounds of air

Answer: 9560 ft³ = 734 pounds of air.

123. Calculate the gas density from an Orsat analysis of the gas with moisture content of 20% (from wet and dry bulb temperatures): $CO_2 = 10.5\%$, $CO = 6.2\%$, $O_2 = 3.0\%$, $N_2 = 80.3\%$. The Orsat analyses are on a dry gas basis.

H_2O = 0.20 × 1.00 × 18	=	3.60
CO_2 = 0.80 × 0.105 × 44	=	3.70
CO = 0.80 × 0.062 × 28	=	1.39

O_2 = 0.80 × 0.03 × 32   = 0.77
N_2 = 0.80 × 0.803 × 28 = 17.99
                                                27.45

gas density = (27.45/28.966) = 0.947 (28.966 is the "apparent" molecular weight of air).

Answer: 0.947 (air = 1.000).

124. A correction factor for excess air is often required in combustion processes. The flue (exhaust) gases are analyzed for CO_2, O_2, and CO with an Orsat apparatus. Nitrogen gas is determined by difference. A flue gas analyzed by an Orsat device contained 10.1% CO_2, 11.1% O_2, and 0.8% CO. Based on an allowable 50% excess air, correct the flue gas dust loading. The flue gas contained 0.493 lb of dust per 1000 lb of gas.

$$\text{ratio of actual : theoretical air} = \frac{N_2}{N_2 - 3.782\left(O_2 - 1/2\,CO\right)} =$$

$$\frac{N_2}{N_2 - 3.782\left(O_2 - 1/2\,CO\right)} = 0.209$$

% total air = 100% + percent excess air

ratio of actual to theoretical air = percent total air/100

$$\frac{(2.09)(0.493\ \text{lb dust}/1000\ \text{lb gas})}{150/100} = 0.687\ \text{lb dust}/1000\ \text{lb gas}$$

Answer: 0.687 pound of dust per 1000 pounds of flue gas.

125. An industrial hygienist would like to estimate a welder's actual exposure to welding fume during the "arc time" discounting interval periods when the welding arc fume generation does not occur. He obtains a three-hr and 36-min air sample at an average air flow rate of 2.13 L/min. The difference in the filter's weight after air sampling was 6.77 milligrams. If seven measurements of the actual electric arc time on this production welder were 21, 19, 14, 13, 26, 20, and 15 sec of arc time per minute, what was the approximate average welding fume concentration during the actual welding process?

$$\frac{21+19+14+13+26+20+15}{7} = \frac{18.3\ \text{seconds}}{60\ \text{seconds}} \times 100 = 30.5\%\ \text{"arc time"}$$

3 hrs and 36 min = 216 min

216 min × (2.13 L/min) = 460 L

$$\frac{6.77 \text{ milligrams}}{0.460 \text{ M}^3} = \frac{14.71 \text{ mg total fumes}}{\text{M}^3} = \frac{1}{0.305} \times \frac{14.71 \text{ mg}}{\text{M}^3} = \frac{48.33 \text{ mg fume}}{\text{M}^3}$$

Answers: 30.5% actual arc time. TWAE level during sample period was 14.71 mg/M³. The average concentration during the welding arc generation was 48.33 mg/M³ which, of course, must not be construed as the welder's true time-weighted average exposure.

126. The relative evaporation rate of toluene has been reported by the American Alliance of Insurers (*Handbook of Organic Solvents*) to be 4.5 times slower than diethyl ether. If 25 mL of "ether" completely evaporates from a flat surface in 21 sec, how long before one gallon of toluene evaporates at similar conditions of liquid and surface temperatures, air temperatures, exposed surface area, and air flow rate over the liquid surface?

"ether": 25 mL/25 sec = average of 1.19 mL evaporates/second

toluene: (1.19 mL/second)/4.5 = average of 0.264 mL evaporates/second

$$1 \text{ gallon} \times \frac{3785 \text{ mL}}{\text{gallon}} \times \frac{\text{second}}{0.264 \text{ mL}} = 14{,}337 \text{ seconds} = 240 \text{ minutes} = 4.0 \text{ hours}$$

Answer: Four hrs (very approximate).

127. A 43-mm long nitrogen dioxide gas permeation tube is used to calibrate a direct reading air sampling instrument. What is the outlet concentration of NO_2 in ppm if the flow rate of NO and NO_2-free nitrogen over the tube is 43 mL/min, and the diluent pure air flow rate is 11.6 L/min? The system temperature is maintained at 30°C, and the permeation rate, PR, for NO_2 gas at this temperature is 1200 nanograms/min-centimeter. The permeation K value (diverse density in L/g) for NO_2 at 30°C is 0.541.

$$43 \text{ mm} = 4.3 \text{ cm} \qquad \frac{1200 \text{ ng}}{\text{minute-cm}} \times 4.3 \text{ cm} = \frac{5160 \text{ nanograms}}{\text{minute}}$$

$$\text{ppm} = \frac{\text{PRxK}}{A+B}, \text{ where:}$$

P R = generation rate of permeation tube, micrograms/min
K = generation rate constant supplied by the manufacturer
A = flow rate of diluent air, L/min
B = flow rate of diluent nitrogen, L/min

$$\text{ppm NO}_2 = \frac{\dfrac{5.16 \text{ mcg}}{\text{min}} \times 0.541}{\dfrac{11.6 \text{ L}}{\text{min}} + \dfrac{0.043 \text{ L}}{\text{min}}} = 0.239$$

Answer: 0.24 ppm NO_2 = 240 ppb NO_2.

128. A large activated charcoal air sampling tube has a desorption efficiency of 68% for an alcohol vapor. The weight of this alcohol reported by an industrial hygiene chemist is 4.23 milligrams for a 119 min air sample obtained at an average air flow rate of 770 mL/min. What is the concentration in ppm if the molecular weight of the alcohol is 74? Assume that the air sampling temperature was 77°F and the barometric pressure was 760 mm Hg.

$$\frac{\text{corrected milligrams}}{\text{sample}} = \frac{\text{detected weight, milligrams}}{\text{desorption efficiency}} = \frac{4.23 \text{ mg}}{0.68} = 6.22 \text{ mg}$$

air volume sampled = 199 min $\times$ 0.77 L/min = 91.63 L

$$\text{ppm} = \frac{\dfrac{6220 \text{ mcg}}{91.63 \text{ L}} \times 24.45}{74} = 22.43 \text{ ppm R} - \text{OH}$$

Answer: 22.4 ppm of alcohol vapor. MeOH, especially, is poorly collected on, and desorbed, from charcoal. Silica gel or other adsorbents must be used.

129. A critical air sampling orifice was calibrated at an elevation of 2000 ft at 65°F and at 1.40 L/min. The orifice will be used at 9000 ft at 80°F. What is the air flow meter correction?

2000 ft = 13.7 psia = 27.82-in Hg   65°F = 18.33°C = 291.3 K
9000 ft = 10.5 psia = 21.39-in Hg   80°F = 26.67°C = 299.7 K

$$\text{Lpm}_{\text{actual}} = \text{LPM}_{\text{indicated}} \times \sqrt{\frac{\text{CP} \times \text{AT}}{\text{AP} \times \text{CT}}}, \text{ where:}$$

P and T are the pressure and temperature in absolute units

C = calibration conditions

A = actual sampling and critical orifice use conditions

$$\text{Lpm}_{\text{actual}} = 1.4 \text{ Lpm} \times \sqrt{\frac{13.7 \text{ psia} \times 299.7 \text{ K}}{10.5 \text{ psia} \times 291.3 \text{ K}}} = 1.62 \text{ L/min}$$

Answer: 1.62 L/min. Correction factor = $\dfrac{1.62 \text{ lpm}}{1.40 \text{ lpm}} = 1.157$. See problem 311 for an explanation of the square root function for orifice air flow meters such as rotameters and critical orifices. Do not use Charles' and Boyle's laws for correction of temperature and pressure for orifice air flow meters.

130. One milligram of typical mineral dust is equivalent to about 30 to 50 million dust particles as determined by standard impinger counting techniques. In a respirable dust cyclone breathing zone air sample obtained at 1.67 L/min for 450 min, the filter weight gain was 0.35 milligrams. What is the weight of

respirable dust per cubic meter? How many particles does this represent? What is this in mppcf?

$(1.67 \text{ L/min}) \times 450 \text{ min} = 751.5 \text{ L}$

$$\frac{350 \text{ micrograms}}{751.5 \text{ liters}} \times \frac{30 \text{ to } 50 \times 10^6 \text{ particles}}{\text{mg dust}} = \frac{14 \text{ to } 23 \times 10^6 \text{ particles}}{M^3} \times$$

$$\frac{M^3}{35.3 \text{ ft}^3} = 0.40 \text{ to } 0.65 \text{ mppcf}$$

Answers: 0.466 mg/M³. 14 – 23 million. 0.40 – 0.65 mppcf.

131. NIOSH reports a coefficient of analysis variation of 0.06 for 1,1-dichloro-ethane. The coefficient of variation for rotameter measurements is 0.05 when air sampling with activated charcoal tubes. What is the total sampling and analysis coefficient of variation?

$$CV = \sqrt{(0.06)^2 + (0.05)^2} = \sqrt{0.0061} = 0.078$$

Answer: The combined sampling and analysis coefficient of variation is 0.078.

132. What volume of air is needed to dilute vapors from one gallon of varnish maker and painter's naphtha below 20% of the lower explosive limit? LEL of VM and P naphtha = 0.9%. Cubic feet of VM and P naphtha per gallon at 70°F = 22.4 ft³. If seven gallons are evaporated from parts every hour in a drying oven, what rate of ventilation (in cfm) is needed to keep VM and P naphtha vapor concentration below 20% of the LEL?

dilution volume (ft³) required for each gallon of solvent =

$$\frac{(100 - \text{LEL})(\text{cubic feet of vapor per gallon})}{20\% \text{ LEL}} = \frac{(100 - 0.9)(22.4 \text{ ft}^3 / \text{gallon})}{0.20 \times 0.9\%} =$$

$$\frac{12,332 \text{ cubic feet of air}}{\text{gallon of VM \& P naphtha}}$$

$$\frac{7 \text{ gallons}}{\text{hour}} \times \frac{12,332 \text{ ft}^3}{\text{gallon}} \times \frac{\text{hour}}{60 \text{ minutes}} = 1439 \text{ ft}^3 \text{ of air}$$

Answers: 12,332 ft³/gallon. 1439 cfm are needed with good, uniform mixing of the dilution air with the vapors and gases in the oven.

133. A 170-cubic inch gasoline engine is running in an enclosed garage at 850 rpm. What volume of exhaust gases is produced every hour? If the exhaust gases

contain 0.76% carbon monoxide by volume, how many cubic feet of CO gas are generated per hour?

$$\frac{\text{engine displacement}\left(in^3\right)\times\text{engine rpm}\times\dfrac{60\text{ minutes}}{\text{hour}}}{2*\times\dfrac{1728\text{ in}^3}{ft^3}}=$$

exhaust volume in ft³/hour (not corrected for temperature of the hot gas)

(*Note:* The denominator of 2 allows for 50% volume of pistons in the engine's cylinders, i.e., some up, some down.)

$$\frac{170\text{ ft}^3\times850\text{ rpm}\times\dfrac{60\text{ minutes}}{\text{hour}}}{2\times\dfrac{1728\text{ in}^3}{ft^3}}=2509\text{ ft}^3/\text{hour}$$

(2509 ft³ total exhaust gases/hr) × 0.0076 = 19.1 ft³ of CO/hr

Answers: 2509 ft³ of total exhaust gases per hour. This engine generates 19.1 ft³ of carbon monoxide gas per hour.

134. If the garage in problem 133 has natural ventilation air exchange rate of 0.5/hr, what is the CO concentration five min after starting the engine? The garage is 4000 ft³.

Contamination of air in enclosed spaces is calculated from the rate of generation of the air pollutant and ventilation of the space [assumes contaminant generation is steady and the ventilation provides uniform air mixing (increase and decay kinetics)].

$$C=\frac{100\,K\left(1-e^{-Rt}\right)}{RV},\text{ where:}$$

C = % (volume/volume) of the gas or vapor in the space after time, t
R = air changes of the space/hr
t = time, hrs
V = volume of the space, ft³
K = contaminant generation rate, ft³/hr

$$C=\frac{100\left[\dfrac{19.1\text{ ft}^3}{\text{hour}}\right]\left(1-e^{-(0.5)(0.083)}\right)}{(0.5)\left(4000\text{ ft}^3\right)}=\frac{\left(1910\text{ ft}^3\right)\left(1-e^{-(0.0415)}\right)}{2000\text{ ft}^3}=$$

$$\frac{(1910)(1-0.959)}{2000}=0.0392\%\text{ CO}=392\text{ ppm CO}$$

Answer: 392 ppm CO. The gas concentration will increase and eventually will plateau as the engine continues to operate and the generation rate is balanced by ventilation loss. 392 ppm CO does not exceed the NIOSH IDLH level of 1500 ppm, but exceeds OSHA's 200 ppm ceiling concentration. Stop the engine. Ventilate.

135. A calibrated length of stain carbon monoxide detector tube indicates the breathing zone concentration of 100 ppm $\pm$ 25% at a barometric pressure of 625 mm Hg. What is the CO gas concentration corrected to sea level?

$$100 \text{ ppm} \pm 25\% \times \frac{760 \text{ mm Hg}}{625 \text{ mm Hg}} = 122 \text{ ppm} \pm 25\% = 90 - 153 \text{ ppm CO}$$

Answer: 90 to 153 ppm carbon monoxide gas.

136. What is the resultant density of air which contains 1% (volume/volume) carbon monoxide gas? Density of 100% CO gas = 0.97 and 100% air = 1.00.

1% CO (v/v) = 10,000 ppm CO in 990,000 ppm air

air: 0.99 $\times$ 1.00   = 0.9900
CO: 0.01 $\times$ 0.97   = 0.0097
resultant density   = 0.9997

Answer: The density of a 10,000 ppm concentration of carbon monoxide gas in air is 0.9997 (essentially equal to that of air). CO gas stratification will not occur.

137. A breathing zone air sample was obtained from a paint sprayer for 443 min at an average rate of 0.83 liter/min using a large charcoal tube preceded by a 37 mm PVC membrane filter cassette. The difference in filter weight after sampling was 6.37 milligrams. The charcoal tube contained 2.97 mg *n*-butyl alcohol, 14.66 mg toluene, 48.49 mg mineral spirits, and 7.01 mg xylene. What was the painter's 8-hr TWAE to these vapor air contaminants? Approximate molecular weights = 74.1, 92.1, 99, and 106.2, respectively. What was the painter's exposure to airborne particulates? Assume an eight-hr exposure.

$$443 \text{ minutes} \times (0.83 \text{ L/minute}) = 367.7 \text{ liters} \qquad \text{ppm} = \frac{\dfrac{\text{micrograms}}{\text{liter}} \times 24.45}{\text{molecular weight}}$$

$$n\text{-butanol:} \quad \frac{\dfrac{2970}{367.7} \times 24.45}{74.1} = 2.7 \text{ ppm} \qquad \text{toluene:} \quad \frac{\dfrac{14,660}{367.7} \times 24.45}{92.1} = 10.6 \text{ ppm}$$

mineral spirits: $\dfrac{\dfrac{48,490}{367.7} \times 24.45}{99} = 32.6 \text{ ppm}$ xylene: $\dfrac{\dfrac{7,010}{367.7} \times 24.45}{106.2} = 4.4 \text{ ppm}$

6370 micrograms/367.7 L = 17.32 mg/M³

$\dfrac{480 \text{ minutes/8 hour work shift}}{443 \text{ minutes}} = 1.084 \text{ multiplier factor}$

Answers: 2.9 ppm *n*-butanol, 11.5 ppm toluene, 35.3 ppm mineral spirits, and 4.8 ppm xylene. 18.77 mg/M³ total airborne particulates. Particulate exposure exceeds the PEL. See problem 138 for solvent vapor TWAEs.

138. What is the additive exposure to solvent vapors in problem 137? Does it exceed the PEL for the mixture? Permissible exposure limits are 50 (C), 100, 100, and 100 ppm, respectively. How many parts of vapor per million parts of air are in the mixture?

2.9 ppm + 11.5 ppm + 35.3 ppm + 4.8 ppm = 54.5 ppm

$\dfrac{2.9}{50} + \dfrac{11.5}{100} + \dfrac{35.3}{100} + \dfrac{4.8}{100} = 0.574 \text{ (57.4\% of the PEL for the mixture}$

There is compliance with the PEL, however the action level is exceeded. It is, therefore, probable that the painter's exposure exceeds the PEL on other days. Even though the solvent vapor exposures may not routinely exceed the PEL, the inhalation of airborne paint solids is excessive. A good organic vapor and paint spray mist respirator or, better, an air-line mask is needed until the ventilation, work practices, and other industrial hygiene controls are substantially improved.

"Nuisance" particulates (not otherwise classified, i.e., there are no compounds of lead, chromates, isocyanates, epoxy resins, etc. in the paint) have an eight-hr TWAE PEL of 10 mg/M³.

Answers: 54.5 total ppm. 57.4% of PEL for the additive vapor mixture. 188% of the "nuisance" particulate limit assuming the paint mist contains no hazardous components.

139. A high volume air sampler ran for eight hrs and seven min at an average air flow rate of 47.3 ft³/min. How many cubic meters of air were sampled?

$\dfrac{\dfrac{1.633 \times 10^9 \text{ mg Cl}_2}{1.376 \times 10^8 \text{ M}^3} \times \dfrac{24.45 \text{ L}}{\text{g-mole}}}{70.9} = 4.1 \text{ ppm Cl}_2$

480 min + 7 min = 487 min

$$487 \text{ min} \times (47.3 \text{ ft}^3/\text{min}) = 23{,}035.1 \text{ ft}^3 \quad \frac{23{,}035 \text{ ft}^3}{35.315 \text{ ft}^3/\text{M}^3} = 652.3 \text{ M}^3$$

Answer: 652.3 cubic meters of air were sampled.

140. 172 g of liquid phosgene splash on a floor. What gas volume quickly results after evaporation at an air temperature of 23°C and an atmospheric pressure of 742 mm Hg? The boiling point of phosgene = 47°F.

$COCl_2$ molecular weight = 12 + 16 + (2 × 35.5) = 99

$$\text{PV=nRT} \qquad n = \frac{172 \text{ grams}}{99 \text{ grams/gram-mole}} = 1.737 \text{ gram-moles}$$

$T = 23°C + 273 = 296 \text{ kelvin}$

$$V = \frac{nRT}{P} = \frac{(1.737 \text{ gram-moles})(0.0821 \text{ L-atm/K-mole})(296 \text{ K})}{\dfrac{742 \text{ mm Hg}}{760 \text{ mm Hg}} = 0.976 \text{ atmosphere}} = 43.183$$

Answer: 43.2 L of phosgene gas. 43.2 × 10^8 L of air would be needed to dilute this gas to 10% of the TLV of 0.1 ppm (dilution to 10 ppb).

141. A detection limit of 40 micrograms/M^3 has been reported for 1, 2-dichloroethane. How long must one sample at 100 mL/min with a small charcoal tube to detect this concentration? The analytical sensitivity for DCE is one microgram.

$$\text{sampling time, minutes} = \frac{1 \text{ mcg}}{\dfrac{40 \text{ mcg}}{10^6 \text{ mL}} \times \dfrac{100 \text{ mL}}{\text{minute}}} = 250 \text{ minutes}$$

Answer: Sample for 250 min (four hrs and 10 min).

142. An electroplater was exposed to HCl gas at four operations: parts dipping (1.6 ppm for 2.5 hrs), parts draining (2.9 ppm for 3.25 hrs), acid replenishment (8.8 ppm for 15 min), and cleanup (0.7 ppm for 0.5 hrs). Assuming that he had negligible exposure for the balance of his eight-hr work shift, what was his TWAE to hydrogen chloride gas?

$C \times T$ = concentration × exposure time = CT = dose (Haber's law)

dipping	1.6 ppm × 2.5 hrs	=	4.00 ppm-hrs
draining	2.9 ppm × 3.25 hrs	=	9.43 ppm-hrs
replenishment	8.8 ppm × 0.25 hr	=	2.20 ppm-hrs
cleanup	0.7 ppm × 0.5 hr	=	0.35 ppm-hr
other	0.0 ppm × 1.5 hrs	=	0.00 ppm-hrs
	8.0 hrs		15.98 ppm-hrs

total dose = (15.98 ppm hrs/8 hrs) = 2.0 ppm TWAE to HCl gas

Answer: An eight-hr time-weighted average exposure to HCl gas of two ppm. His exposure when replenishing acid exceeds the five ppm OSHA STEL. Perhaps a full-face acid-gas respirator and improved ventilation should be implemented at this operation.

143. A 4-1/2 pound chunk of calcium phosphide fell into a vat of water and released phosphine gas according to the following reaction. How much phosphine gas was generated disregarding water solubility and assuming stoichiometric conversion?

$$Ca_3P_2 + 6\ H_2O \rightarrow 3\ Ca(OH)_2 + 2\ PH_3\uparrow$$

molecular weight of $Ca_3P_2 = 182.19$

molecular weight of PH_3 gas $= 34.00$

4.5 lb $Ca_3P_2 \times 454$ g/lb $= 2043$ g Ca_3P_2

$$\frac{2043\ \text{grams}\ Ca_3P_2}{182.19\ \text{grams/mole}} = 11.21\ \text{moles of}\ Ca_3P_2$$

11.21 mols of $Ca_3P_2 \rightarrow 22.42$ mols of $PH_3\uparrow$

22.42 mols $\times$ 34.00 g/mol $= 762.28$ g PH_3

Answer: 762 g of phosphine gas were released.

144. In problem 143, what would be the phosphine gas concentration in ppm if this occurred in an unventilated room with dimensions of $20 \times 50 \times 50$ ft? Assume a uniform mixing of the PH_3 gas with the room air.

$20 \times 50 \times 50$ ft $= 50,000$ ft³ $= 1416$ M³

$$ppm = \frac{\dfrac{762,280\ \text{mg}}{1416\ \text{M}^3} \times 24.45}{34} = 387\ \text{ppm}\ PH_3$$

Answer: 387 ppm PH_3. TLV and PEL of phosphine gas are only 0.3 ppm with a STEL of 1 ppm. SCBAs, training, ventilation, etc. are necessary.

145. ASHRAE recommends dust collectors, not air filters, to clean exhaust air when the dust concentrations exceed four grains of dust per one thousand cubic feet of exhaust air. What is this dust concentration in mg/M³?

$$\frac{4\ \text{grains}}{1000\ \text{ft}^3} \times \frac{35.315\ \text{ft}^3}{\text{M}^3} \times \frac{0.06480\ \text{gram}}{\text{grain}} \times \frac{1000\ \text{mg}}{\text{gram}} = \frac{9.15\ \text{mg}}{\text{M}^3}$$

Answer: 9.15 milligrams of total particulates per cubic meter of air.

146. A vertical hazardous liquid waste incinerator has a gas flow rate of 6900 cfm at 70°F. The interior dimensions of the incinerator are nine ft square and 37 ft high. The design operating temperature of the incinerator is 2200°F. What is the residence time for a molecule of vapor in the incinerator? What is the actual flow rate of the gas? What incinerator dimensions are required for a three-second residence time?

70°F = 530°R   2200°F + 460 = 2660°R

applying Charles' law: $\dfrac{V_i}{V_f} = \dfrac{T_i}{T_f}$:

$$V_f = \frac{T_f \times V_i}{T_i} = \frac{6900 \text{ cfm} \times 2660°R}{530°R} = 34{,}630 \text{ cfm} \qquad Q = AV$$

$$V = Q/A = \frac{34{,}630 \text{ cfm}}{9 \times 9 \text{ ft}} = 428 \text{ fpm} \qquad \frac{428 \text{ fpm}}{60 \text{ seconds/minute}} = 7.1 \text{ feet/second}$$

$$\frac{37 \text{ feet}}{7.1 \text{ feet/second}} = 5.2 \text{ seconds average residence time/molecule}$$

incinerator volume = 9 × 9 × 37 ft = 2997 ft³

incinerator volume for a three-second residence time =

$$\frac{2997 \text{ ft}^3}{5.2 \text{ seconds}} = \frac{x \text{ ft}^3}{3 \text{ seconds}}$$

x = 1729 ft³. The length to cross-sectional area ratios must remain identical to ensure necessary transit velocity and molecule residence time in the combustion space.

Answers: 5.2 sec. 34,630 cfm. 1727 ft³. Perhaps a smaller incinerator could be used, say, 7 × 7 × 37 ft which would provide a 3.1 second residence time.

147. Industrial waste liquid chlorobenzene is fed into a large vertical hazardous waste incinerator at an atomizer nozzle feed rate of 1670 lb/hr. What are the stack gas combustion products assuming 100% oxidation with 70% excess air?

$$C_6H_5Cl + 7 O_2\left[\frac{0.79}{0.21} \times 7 N_2\right] \rightarrow HCl \uparrow + 6 CO_2 \uparrow + 2 H_2O \uparrow + \left[\frac{0.79}{0.21} \times 7 N_2\right]$$

at 70% excess air:

$$C_6H_5Cl + (1.7 \times 7)O_2 + \left[1.7 \times \frac{0.79}{0.21} \times N_2\right] \rightarrow 6 CO_2 \uparrow + 2 H_2O \uparrow + HCl \uparrow$$

$$(0.6)\, 7\, O_2 + 7\left[1.6 \times \frac{0.79}{0.21}\right] N_2$$

molecular weights: chlorobenzene = 112.5, oxygen = 32, nitrogen = 28, HCl = 36.5, carbon dioxide = 44

one pound-mol of chlorobenzene requires seven pound-mols of O_2

$$\frac{1670 \text{ pounds/hour}}{112.5 \text{ lb/lb-mole}} = \frac{7 \times 1670 \text{ pounds-moles } O_2}{112.5} \qquad 1670 / 112.5 = 14.84$$

CO_2: $\dfrac{1670}{112.5} \times 6 = 89.07$ H_2O: $\dfrac{1670}{112.5} \times 2 = 29.69$, etc.

incinerator combustion products:

CO_2:	6 lb-mol × 44 × 14.84	=	3918 lb
H_2O:	2-lb-mol × 18 × 14.84	=	534 lb
HCl:	1 lb-mol × 36.5 × 14.84	=	542 lb
O_2:	0.6 lb-mol × 7 × 32 × 14.84	=	1995 lb

N_2: $\dfrac{1.6 \times 0.79}{0.21}$ lb-mole × 7 × 28 × 14.84 = 17,507 lb

total = 24,496 lb

3918 lb/24,496 lb   = 0.160
534 lb/24,496 lb   = 0.022
542 lb/24,496 lb   = 0.022
1995 lb/24,496 lb   = 0.081
17,507 lb/24,496 lb = 0.715
                      1.000

Answers: 24,500 lb/hr. 16% CO_2, 2.2% H_2O, 2.2% HCl, 8.1% O_2, and 71.5% nitrogen-argon.

148. A solvent paint stripper is 30% by volume methylene chloride and 70% by volume methanol. What is the volume percent composition of the vapor at normal room temperature? The vapor pressures of CH_2Cl_2 and MeOH are 350 mm Hg and 92 mm Hg, respectively. Their respective densities are 1.33 g/mL and 0.79 g/mL, and their molecular weights are 84.9 and 32.1, respectively.

using 100 mL of the solvent mixture as a basis for Raoult's law calculations:

30 mL × 1.33 g/mL = 39.9 g CH_2Cl_2
70 mL × 0.79 g/mL = 55.3 g MeOH

$$\text{partial pressure CH}_2\text{Cl}_2 = \frac{\dfrac{39.9\,g}{84.9\,g/mole} \times 350\,mm\,Hg}{\dfrac{39.9\,g}{84.9\,g/mole} + \dfrac{55.3\,g}{32.1\,g/mole}} = 75\,mm\,Hg$$

$$\text{partial pressure MeOH} = \frac{\dfrac{55.3\,g}{32.1\,g/mole} \times 92\,mm\,Hg}{\dfrac{39.9\,g}{84.9\,g/mole} + \dfrac{55.3\,g}{32.1\,g/mole}} = 72.3\,mm\,Hg$$

75 mm Hg + 72.3 mm Hg = 147.3 mm Hg total vapor pressure for both solvents

$$\frac{75\,mm\,Hg}{147.3\,mm\,Hg} \times 100 = 50.9\% \text{ methylene chloride vapor}$$

$$\frac{72.3\,mm\,Hg}{147.3\,mm\,Hg} \times 100 = 49.1\% \text{ methyl alcohol vapor}$$

Answers: 50.9% CH_2Cl_2 and 49.1% MeOH in the vapor phase. Refer to problem 18 for a discussion of the application and deviations from Raoult's law.

149. A rotameter calibrated in cubic feet per hour was used in an air sampling train operating for two hrs and 39 min. What volume of air was sampled in liters if the average rotameter reading was 4.6?

$$\frac{4.6\,ft^3}{hour} \times \frac{hour}{60\,minutes} \times 159\,minutes \times \frac{28.32}{ft^3} = 345\,L$$

Answer: 345 L of air.

150. A solvent drum filling operator has an average exposure to 83 ppm methylene chloride for 3 1/2 hrs, 32 ppm isopropyl alcohol for 1 1/2 hrs, and 27 ppm toluene for three hrs. If their respective TLVs are 50 ppm, 400 ppm, and 100 ppm, are solvent vapor exposure controls warranted?

$$\frac{83\,ppm \times 3.5\,hours}{50\,ppm \times 8\,hours} \times 100 = 72.6\% \text{ of the TLV for } CH_2Cl_2$$

$$\frac{12\,ppm \times 1.5\,hours}{100\,ppm \times 8\,hours} \times 100 = 1.5\% \text{ of the TLV for IPA}$$

$$\frac{27\,ppm \times 3\,hours}{100\,ppm \times 8\,hours} \times 100 = 10.1\% \text{ of the TLV for toluene}$$

72.6% + 1.5% + 10.1% = 84.2% of the additive TLV for the work shift exposure

Answer: Yes, industrial hygiene controls are required because the combined exposure exceeds the action level based upon the additive toxicological effects. **Remember:** *Permissible exposure limits are the worst acceptable conditions, and must never be regarded as goals, objectives, and end points.* A major industrial hygiene premise is the control of every exposure to as low as possible using the best available technology. Control of methylene chloride vapors is especially important since it is a possible human carcinogen and has multiple toxic effects, and contributes the highest percentage exposure to the mixture of vapors.

151. A railroad hopper car filler has an eight-hr time-weighted average exposure to mixed grain dust (barley, oats, wheat) of 3.4 mg/M³ (TLV = 4 mg/M³). He has a simultaneous exposure to respirable silica (α-quartz) released from the cascading dry grain of 0.036 mg SiO_2/M³ (TLV = 0.1 mg respirable quartz/M³). Are industrial hygiene controls needed? If so, what would you prescribe?

$$\frac{3.4 \text{ mg/M}^3}{4 \text{ mg/M}^3} + \frac{0.036 \text{ mg/M}^3}{0.1 \text{ mg/M}^3} = 0.85 + 0.36 = 1.21 \qquad 21\% > TLV_{mixture}$$

Answer: Industrial hygiene controls (powered air-purifying respirator?, working inside a filtered air booth?, PFTs, training, etc.) are necessary since exposure to grain dust exceeds the action level, and the combined exposure to both dusts is 2.4 times the action level. Since adverse pulmonary effects (fibrosis) are common to both dusts, good control of inhalation exposures assumes great importance.

152. A calibrated combustible gas indicator was used to measure explosive vapors in a gasoline tank. A gas/vapor sampling probe attached to the CGI was lowered into the tank. The needle immediately "pegged" above 100% of the LEL and then quickly dropped to less than 2% of the LEL. Assuming a nonabsorptive sampling gas/vapor sampling probe, what should you conclude?

a. The meter correctly indicates less than 2% of the LEL.
b. A CGI Wheatstone bridge electronic circuit is quickly "poisoned" by organic lead vapors emitted from the gasoline and then reads low.
c. There is less than 4% oxygen in the tank's vapor space.
d. Since gasoline is a mixture of numerous hydrocarbons, an inaccurate reading results if the instrument was not calibrated with the actual gasoline vapor being tested.
e. The vapor concentration is not less than 2% of the LEL; it is actually above the UEL.

Answer: e. The concentration of vapor is so rich that no combustion can occur in the instrument. This is a potentially dangerous situation because unsuspecting persons may assume a safe atmosphere with respect to explosion. As the vapors are diluted down into the LEL-UEL range with air, an explosion will occur if there is an ignition source. Dilute the explosive vapors with nitrogen to < 20% LEL, then use air to < TLV and PEL. Verify concentrations of oxygen, flammable vapors, and toxics with recently-calibrated combustible gas indicator, oxygen level meter, and atmospheric toxicants instruments.

153. A paint sprayer has eight-hr time-weighted average exposures to toluene at 79 ppm (TLV = 100 ppm), n-butyl alcohol at 16 ppm ("C" = 50 ppm/Skin), and noise at 84 dBA. What are the major industrial hygiene issues?

Answer: In addition to the additive narcotic effects of the two solvent vapors, reports in occupational medical literature have indicated loss of hearing in people with chronic work place exposures to these solvents. Therefore, effects of the inhaled solvent vapors are presumed additive to the physical effects of noise on the painter's hearing. All elements of a hearing conservation program must be applied along with careful steps to reduce vapor and skin contact exposures.

154. A maintenance welder is exposed to 3.6 mg iron oxides/M^3 and 27.4 mg of total dust per M^3 in a corn flour mill. What should you recommend to protect his health from these dust and fume exposures?

Answer: First, get him to **stop welding** before he blows the mill and everybody who works there to smithereens. Start a fire safety and welding safety training program including "hot work" permits, grain dust explosion prevention practices, ventilation engineering, respiratory protection, supervision, plant housekeeping, etc.

155. How much air is exhausted from a drying oven operating at 320°F if 1360 cubic feet of air at 65°F are supplied per minute to the oven?

65°F + 460°F = 525° absolute (Rankine)
320°F + 460°F = 780° absolute (Rankine)

$$\frac{780° \text{ R}}{525° \text{ R}} \times \frac{1360 \text{ ft}^3}{\text{minute}} = \frac{2020.6 \text{ ft}^3}{\text{minute}}$$

Answer: 2021 cfm at 320°F, although the mass of hot air remains the same.

156. A gray iron foundry cupola attendant leaves work after six hrs going home with an intense headache, dizziness, and nausea. This 26-year-old man, feeling better after an hour, decides to strip paint from an old chair in his basement. His wife finds him an hour later collapsed on their basement floor. He died

from cardiac arrest after ventricular fibrillation. His medical history includes chronic anemia. What is a reasonable forensic conclusion?

a. He should not have gone to work that morning.
b. There was an oxygen-deficient atmosphere near the cupola.
c. Silica dust and formaldehyde gas in the foundry act in an additive manner with methylene chloride vapors which volatilize from paint strippers.
d. He should have worn a good dust respirator while at work so his lungs would be capable of detoxifying any solvent vapors he inhaled while working in his basement.
e. The carbon monoxide he inhaled at work was additive to the *in vivo* conversion of inhaled methylene chloride vapor to carbon monoxide while at home.
f. Anemia was a significant risk factor.

Answer: e. Cupola workers have potential exposures to high concentrations of CO gas. His symptoms were consistent with a COHb concentration exceeding 10–20% or more. His symptoms occurred late in the work shift which indicates that he went home with a significant body burden of COHb. Methylene chloride vapor, in addition to being a potential sensitizer of the myocardium, is present in most paint strippers and metabolizes partly *in vivo* to carbon monoxide.

157. Air was sampled at an average flow rate of 2.14 L/m through a 37 mm membrane filter for seven hrs and 47 min. The sample was obtained to measure a welder's TWAE to aluminum fume as the oxide, Al_2O_3. An industrial hygiene chemist analyzed 7.93 milligrams of aluminum on the filter after blank correction. What was the concentration of aluminum oxide fume? The atomic weight of aluminum = 27, the molecular weight of aluminum oxide = 102.)

$$\frac{1 \text{ aluminum oxide}}{2 \text{ aluminum}} = \frac{102}{2 \times 27} = 1.89 = \text{ the Al to } Al_2O_3 \text{ molar conversion factor}$$

$$7 \text{ hrs, } 47 \text{ min} = 467 \text{ min} = \frac{467}{480} \times 100 = 97.3\% \text{ of an 8-hr work shift}$$

$$467 \text{ min} \times 2.14 \text{ L/m} = 999.4 \text{ L}$$

$$\frac{7930 \text{ micrograms aluminum}}{999.4 \text{ L}} = 7.93 \text{ mg Al/M}^3$$

$7.93 \text{ mg Al/M}^3 \times 1.89 = 14.99 \text{ mg } Al_2O_3/M^3 \text{ TWAE}$
$14.99 \text{ mg/M}^3 \times 467 \text{ min} = 7000.33 \text{ mg/M}^3\text{-min}$
$0 \text{ mg/M}^3 \times 13 \text{ min} = 0 \text{ mg/M}^3\text{-min}$

$$\frac{7000.33 \text{ mg/M}^3 - \text{minutes}}{480 \text{ minutes}} = 14.58 \text{ mg/M}^3 \text{ TWAE}$$

(assuming that the balance of his/her work shift is free of Al_2O_3 fume exposure)

Answer: 14.58 mg of aluminum oxide per cubic meter. This is anomalous because the TLV for aluminum oxide is 10 mg/M³ (as aluminum). This exposure to the fume meets the TLV when calculated as aluminum, but exceeds the TLV and PEL of 10 mg/M³ for nuisance particulates. Clearly, however, this exposure to welding fume requires better control. Since welding of aluminum is often done by MIG technique, ozone gas can be generated in large amounts. *Note:* There is a specific TLV of five mg/M³ for aluminum welding fume which resolves the apparent anomaly. Al_2O_3 dust has a justifiably higher TLV than that for Al_2O_3 fumes.

158. Convert 1.78 mg hydrogen fluoride gas per cubic meter of air to ppm. Molecular weight of HF = 20.

$$ppm = \frac{\frac{mg}{M^3} \times 24.45}{molecular\ weight} = \frac{\frac{1.78\ mg}{M^3} \times 24.45}{20} = 2.18\ ppm\ HF\ gas$$

Answer: 2.2 ppm HF gas. HF gas ceiling concentration = 5 ppm.

159. An industrial hygiene chemist detected less than five micrograms of toluene in a charcoal tube used in the collection of a 9.76 liter air sample. What was the concentration of toluene vapor?

$$\frac{\leq 5\ mcg\ toluene}{9.76\ L} \leq 0.512\ mcg/L \qquad molecular\ weight\ of\ toluene = 92$$

$$\frac{\frac{\leq 0.512\ mcg}{L} \times 24.45}{92} \leq 0.136\ ppm$$

Answer: Less than 0.14 ppm toluene ($\leq$ 140 ppb).

160. What was the average flow rate of an air sampling pump if pre-sampling and post-sampling elapsed times through a 1000 mL soap bubble apparatus were 83.5 and 84.9 sec, respectively?

$$\frac{83.5\ seconds + 84.9\ seconds}{2} = 84.2\ seconds\ average$$

$$\frac{60\ seconds/minute}{84.2\ seconds/liter} = 0.713\ L/m$$

Answer: Average air flow rate = 0.713 L/min = 713 mL/min.

161. What is the density of methyl chloroform vapor? The density of air at NTP = 1.2 milligrams per cubic centimeter = 1.2 mg/mL.

CH_3CCl_3 molecular weight $= (2 \times 12) + (3 \times 35.5) + (3 \times 1) = 133.5$

$$1 \text{ ppm } CH_3CCl_3 = \frac{\text{molecular weight} \times \frac{\text{mcg}}{L}}{24.45} = \frac{133.5}{24.45} = \frac{5.46 \text{ mg}}{M^3} \approx \frac{1 \text{ mL}}{M^3}$$

$$\frac{5.46 \text{ mg MC/mL}}{1.2 \text{ mg air/mL}} = 4.55$$

Answer: Density of 100% methyl chloroform vapor = 4.55 (air = 1.00).

162. Convert one ppm hydrogen selenide gas to mg/M³. The molecular weight of H_2Se is 81.0.

$$1 \text{ ppm } H_2Se = \frac{\frac{\text{mcg}}{L} \times 24.45}{81} = \frac{3.31 \text{ mcg}}{L} = \frac{3.31 \text{ mcg}}{M^3}$$

Answer: 1 ppm H_2Se = 3.31 mg H_2Se/M³.

163. Knowing the concentration of carbon monoxide gas in inhaled air and a few other important parameter variables permits one to estimate the percent unsaturated hemoglobin in a CO-exposed person. An equation for this is:

$$\% \text{ COHb} = \left(2.76 \times e^{\frac{h}{7000}} \right) + \left(0.0107 \text{ a } C^{0.9} t^{0.75} \right), \text{ where:}$$

h = height above sea level (feet), a = activity (3 = rest, 5 = light activity, 8 = light work, and 11 = heavy work), C = ppm inhaled CO, and t = the exposure duration (hours).

What is the predicted COHb concentration in a man performing light work at 8000 ft for six hrs at 50 ppm CO?

$$\% \text{ COHb} = \left(2.76 \times e^{\frac{h}{7000}} \right) + 0.0107 \left(8 \times 50^{0.9} \times 6^{0.75} \right) = \left(2.76 \times e^{1.142} \right) +$$

Answer: About 20% COHb; that is, a worker who has an intense headache, weakness, and perhaps nausea, vomiting, and dimness of vision, tinnitus, etc. and possibly a myocardial infarction if he/she has a history of ischemic heart disease or other compromised cardiac capacity.

164. The "standard man" in toxicology is assumed to weigh 70 kilograms (= 154.35 pounds). He inhales 22,800 L of air per day: 9600 L at eight hrs of light work, 9600 L at eight hrs nonoccupational, and 3600 L resting and sleeping. What

weight of air does he inhale every year? Dry air weighs 1.2 kg/M³ at 21°C
and at sea level. Assume that this "standard man" lives in Seattle.

$$22,800\,L = 22.8\,M^3 \quad \frac{22.8\,M^3}{day} \times \frac{365.25\,days}{year} \times \frac{1.2\,kg}{M^3} = \frac{9993\,kg}{year}$$

$$\frac{9993\,kg}{year} \times \frac{2.205\,lb}{kg} = \frac{22,035\,lb}{year}$$

Answer: About 22,000 pounds (11 tons) per year. Does this mean exhalation
of 11 tons of bad breath per year (halitosis TLV = 1 ppb)?

165. What factors are of minor significance in stack sampling?

   a. air/gas density, condensed water, dust thimble weight
   b. stack gas flow rate, isokinetic sampling, temperature
   c. suction pressure at meter, condensed water, gas flow rate
   d. molecular weight of air, friction losses, season of the year
   e. flow rate through sampling train, barometric pressure, sample duration
   f. Pitot correction factor, gas:dust ratio, presence of reactive gases, temperature

Answer: d.

166. An air pollution chemist wants to determine the emission rate of chlorine gas
from a vent in a chemical synthesis reactor. Which of the following should be
of minor, negligible significance to her?

   a. concentration of chlorine in the gas stream
   b. isokinetic sampling
   c. density of air or gas stream containing the Cl_2 gas
   d. process operation or cycle time
   e. volumetric total air or gas emission rates
   f. moisture content of the air or gas stream

Answer: b. Isokinetic stack sampling is not required when only gases or
vapors are determined. Isokinetic sampling is mandatory for airborne particulates in ducts, stacks, and exhaust vents (e.g., dust, fume, smoke, mist, fibers,
aerosols, spray, etc.). All other parameters are important in her analysis.

167. A 10-in internal diameter stack discharges air containing 23 ppm chlorine gas
at 670 ft/min. What is the chlorine gas mass emission rate?

$$stack\ area = \pi \frac{(5\,in)^2}{\dfrac{144\,in^2}{ft^2}} = 0.5454\,ft^2$$

$$Q = AV = 0.5454\,ft^2 \times 670\,fpm = 365.4\,cfm$$

$$\frac{mg}{M^3} = \frac{ppm \times molecular\ weight}{24.45} = \frac{23 \times 71}{24.45} = \frac{66.79\ mg}{M^3}$$

$$\frac{66.79\ mg}{M^3} \times \frac{M^3}{35.3\ ft^3} \times \frac{365.4\ ft^3}{minute} = \frac{691.4\ mg\ Cl_2}{minute}$$

Answer: 691 milligrams of chlorine are released per minute.

168. The LEL for methyl ethyl ketone at normal room temperature is 1.7%. What is the MEK vapor level expressed in g/M³?

molecular weight of CH_3-CO-CH_2CH_3 = (4 carbon $\times$ 12) + (1 oxygen $\times$ 16) + (8 hydrogen $\times$ 1) = 72

1.7% = 17,000 ppm (vol/vol)

$$\frac{mg}{M^3} = \frac{ppm \times molecular\ weight}{24.45} = \frac{17,000 \times 72}{24.45} = \frac{50,061.4\ mg\ 2\text{-butanone}}{M^3}$$

Answer: MEK LEL = 50 g of vapor per cubic meter of air.

169. The LEL for MEK (2-butanone) is 1.7% (vol/vol) at standard room temperatures. A vapor concentration of 1.4% (vol/vol) in air explodes. What can be concluded?

a. The LEL of 1.7% must be wrong.

b. The 1.4% vapor-in-air mixture must have been hotter than the standard room temperature.

c. The data provided are insufficient to conclude anything.

d. The ignition source for the 1.4% vapor must have been hotter than normal.

e. The oxygen concentration for the 1.7% LEL determination was higher than existed in the explosion of the 1.4% concentration.

Answer: b. Raising the temperature lowers the LEL, e.g., the LEL for MEK is 1.4% for this air : solvent vapor mixture at 200°F.

170. An air pollution control baghouse collects 93 pounds of dust for every eight-hr operating period. The ventilation duct leading to the baghouse plenum has an 18-in internal diameter with an average duct velocity of 3250 ft/min. What is the average airborne dust concentration in the influent duct in mg/M³?

18-in diameter = 1.5-ft diameter   duct radius = 0.75 ft

$$Q = AV = \pi \ (0.75 \times 0.75 \ \text{ft})^2 \times 3250 \ \text{fpm} = 5743 \ \text{cfm}$$

$$\frac{93 \ \text{pounds}}{480 \ \text{pounds}} = \frac{0.194 \ \text{pound}}{\text{minute}} \qquad \frac{\dfrac{0.194 \ \text{lb}}{\text{minute}}}{\dfrac{5743 \ \text{ft}^3}{\text{minute}}} = \frac{0.0000338 \ \text{pound}}{\text{ft}^3}$$

$$\frac{0.0000338 \ \text{lb}}{\text{ft}^3} \times \frac{454 \ \text{g}}{\text{lb}} \times \frac{1000 \ \text{mg}}{\text{gram}} \times \frac{35.3 \ \text{ft}^3}{\text{M}^3} = \frac{542 \ \text{mg}}{\text{M}^3}$$

Answer: Approximate average of 542 milligrams of particulates/cubic meter.

171. A carbon adsorption unit air pollution control device operating at 6600 cfm has an inlet vapor concentration of 360 ppm and an outlet concentration of 25 ppm. What is the annual savings if the unit operates 100 hrs per week and 50 weeks per year? The cost of this solvent is $4.83/gallon. The solvent's molecular weight is 114. The solvent's density is 0.83 g/mL. Assume a 97% recovery of the solvent vapor adsorbed on the charcoal.

360 ppm vapor at inlet – 25 ppm at outlet = 335 ppm collection (a 92.5% efficiency)

$$\frac{\text{mg}}{\text{M}^3} = \frac{\text{ppm} \times \text{molecular weight}}{24.45} = \frac{335 \times 114}{24.45} = \frac{1562 \ \text{mg}}{\text{M}^3}$$

$$\frac{1562 \ \text{mg}}{\text{M}^3} \times \frac{\text{M}^3}{35.3 \ \text{ft}^3} \times \frac{\text{gram}}{1000 \ \text{mg}} \times \frac{\text{pound}}{454 \ \text{grams}} = \frac{0.0000975 \ \text{lb}}{\text{ft}^3}$$

$$\frac{0.0000975 \ \text{lb}}{\text{ft}^3} \times \frac{6600 \ \text{ft}^3}{\text{minute}} \times \frac{60 \ \text{minutes}}{\text{hour}} \times \frac{5000 \ \text{hours}}{\text{year}} \times 0.97 =$$

$$\frac{187{,}259 \ \text{lb solvent recovered}}{\text{year}}$$

$$\frac{187{,}259 \ \text{lb}}{\text{year}} \times \frac{454 \ \text{grams}}{\text{lb}} \times \frac{\text{mL}}{0.83 \ \text{g}} \times \frac{\text{gallon}}{3785 \ \text{mL}} \times \frac{\$4.83}{\text{gallon}} = \frac{\$130{,}708}{\text{year}}$$

Answer: An annual savings of $130,708 minus maintenance costs, labor costs, taxes, operating costs, and capital equipment depreciation. Other opportunities for savings include alternative technologies, improved vapor collection efficiency, improved liquid solvent recovery efficiency, and the use of a less costly solvent.

172. A 210 ft³ compressed gas cylinder containing 10% by volume carbon monoxide in air crashes to the floor in a 10 × 30 × 60-ft room which has no ventilation. There is no protective cover over the gas valve. The valve snaps

off, and the CO gas mixture is quickly released. The depressurizing, rapidly whirling cylinder provides uniform gas mixing in the room. What is the final CO gas concentration after completely mixing with the room air? Disregard the small increase in the overall atmospheric pressure in the room from the released gas.

210 ft³ × 0.10 = 21 cubic feet of carbon monoxide

10 × 30 × 60 ft = 18,000 cubic feet

$$\frac{21 \text{ ft}^3 \text{ of CO}}{18,000 \text{ ft}^3} \times 10^6 = 1167 \text{ ppm CO}$$

Answer: 1167 parts of carbon monoxide gas per million parts of air. It should be noted that this answer is slightly in error because an amount of CO equal to the cylinder volume will remain in the cylinder when it comes to atmospheric pressure. For example, subtract three cubic feet from 210 ft³ if this is the cylinder's volume. See problem 327.

173. What volume of propane gas is required to yield a concentration of 1000 ppm in a gas calibration chamber with an internal volume of 765 L?

$$\text{ppm} = \frac{\text{volume of gas}}{\text{volume of air}} \times 10^6 \qquad \text{volume of gas} = \frac{1000 \times 765 \text{ L}}{10^6} = 0.765 \text{ L}$$

Answer: 0.765 L = 765 mL of propane gas.

174. The average dust concentration in a bag house inlet plenum is 348 mg/M³. The outlet dust concentration is 4.7 mg/M³. What is the average collection efficiency of the bag house filters for this dust?

percent collection efficiency =

$$100 \left[1 - \frac{C_{out}}{C_{in}} \right] = 100 \left[1 - \frac{4.7 \text{ mg/M}^3}{348 \text{ mg/M}^3} \right] = 100 \left(1 - 0.0135 \right) = 98.65\%$$

Answer: 98.65% dust collection efficiency. Please note that this is the average mass collection efficiency. Since particle weights are proportional to the cube of their diameters, the mass collection efficiency does not equal the particle collection efficiency. Said another way, depending upon type of dust, for example, 80% of the particles by weight might only comprise 2% of the particles by count. Recirculating the exhaust air from a dust bag collector with a collection "efficiency" of 95%, e.g., might be very hazardous to workers. Always ask manufacturers of air pollution control devices how they determined the collection efficiency of their products.

175. One g-mol of SO_2 occupies _____ liters at _____ °C and _____ mm
     Hg and contains _____ molecules.

     a. 22.4, 25, 760, 6 × 10^{23}
     b. 24.45, 0, 760, 6 × 10^{21}
     c. 24.45, 25, 745, 6 × 10^{23}
     d. 22.4, 0, 760, 2.023 × 10^{23}
     e. 24.45, 25, 760, 6 × 10^{23}

     Answer: e.

176. One milliliter of methyl chloroform becomes what vapor volume when evap-
     orated at a pressure of 720 mm Hg and a temperature of 25°C?

     $$\text{density} = 1.34 \text{ g/mL} \qquad PV = nRT \qquad \frac{1.34 \text{ gram}}{133.4 \text{ g/mole}} = 0.010 \text{ mole}$$

     $$V = \frac{nRT}{P} = \frac{(0.010 \text{ mole})(0.0821 \text{ atm-liter/mole-K})(298 \text{ K})}{(720 \text{ mm Hg}/760 \text{ mm Hg}) = 0.947 \text{ atmosphere}} = 0.258 \text{ liter}$$

     Answer: 258 mL of 100% methyl chloroform vapor.

177. _____ refers to the number of molecules per cubic centi-
     meter of an ideal, perfect gas at 0°C (2.6782 × 10^{19}).

     a. Avogadro's number
     b. Loschmidt's number
     c. Fanning's friction factor
     d. Boyle's number
     e. Dalton's number

     Answer: b.

178. A 20 × 100 × 120-ft building has a ventilation system providing 6000 cfm
     of outside air. The outdoor design condition is 20°F at 60% relative humidity
     (nine grains of moisture per pound). If we want to maintain 50% relative
     humidity at 75°F (66 grains of moisture per pound of air) at 13.78 cubic feet
     per pound of air in the air conditioned space, how much water must be added?

     $$\text{the formula for humidification load} = \frac{(\text{CFH})(\text{G})}{(\text{V})(7000)}, \text{ where:}$$

     CFH = cubic feet of air per hour (= cfm × 60 min/hr)
     G   = grains of moisture per pound of inside air minus grains per pound
           outside air
     V   = specific volume of inside air in cubic feet per pound of air
     7000 = conversion factor, grains of moisture per pound

$$\frac{(6000 \times 60)(66-9)}{(13.78)(7000)} = 212.7 \text{ pounds of water per hour}$$

$$\frac{212.7 \text{ pounds H}_2\text{O/hour}}{8.33 \text{ pounds/gallon}} = \frac{25.5 \text{ gallons to be evaporated}}{\text{hour}}$$

Answer: About 26 gallons of water per hour. If there was no ventilation system, an air infiltration rate of two air changes/hr could be assumed for a building with reasonably tight construction in a cold climate. Instead of the (6000 × 60) for CFH, we would use 2 (100 × 120 × 20 ft). The humidification load then becomes 285 lb/hr.

179. One milligram of quartz dust in the size range reported for industrial atmospheres typically contains 200 to 300 million particles. What does one hypothetical silica dust particle weigh?

$$\frac{250 \times 10^6 \text{ particles}}{1000 \text{ micrograms}} = \frac{250,000 \text{ particles}}{\text{microgram}} = \frac{250 \text{ particles}}{\text{nanogram}} = \frac{0.25 \text{ particle}}{\text{picogram}}$$

Answer: An average particle weighs about four picograms (4×10^{-12} g).

180. The behavior of airborne particles can best be described by which of the following physical laws and factors:

a. Stokes' law, Archimedes' principle, Newtonian gravity, and Brownian motion
b. Frank's friction factor, Cunningham's factor, gravity, and Newtonian gravity
c. Brownian motion, Avogadro's law, gravity, and ion flux
d. Newtonian gravity, Stokes' law, Drinker's coefficient, and McWelland's aerodynamic slip factors
e. gravity, Stoke's law, Cunningham's factor, and Brownian motion

Answer: e.

181. A granite quarry worker's exposures to dust were 192 mppcf for 3 3/4 hrs at drilling, 1260 mppcf for 15 min blowing out holes, 12 mppcf for 1 1/4 hrs changing drills, 9.8 mppcf for 2 1/4 hrs watching drills, and 8.1 mppcf for 1/2 hr while broaching. What was his eight-hr time-weighted average exposure to airborne dust?

job	mppcf	×	hours	=	mppcf-hours
drilling	192	×	3.75	=	720
blowing out holes	1260	×	0.25	=	315
changing drills	12	×	1.25	=	15
watching drills	9.8	×	2.25	=	22
broaching	8.1	×	0.5	=	4
			8.0		1076

1076 mppcf-hrs/8.0 hrs = 134.5 mppcf (% α- and β-quartz in dust?)

Answer: Eight-hr time-weighted average exposure = 134.5 mppcf. Dusty. How about prescribing a PAPR, wet methods of dust suppression, mechanical local exhaust ventilation systems, and improvements in his work practices?

182. A drying oven vents MEK vapor into a work space. If the solvent consumption rate is steady at one pint every four min, and the concentration of MEK vapor leaving the oven is 200 ppm, what is the ventilation rate of the oven? The vapor volume for MEK = 4.5 cubic feet of vapor per pint evaporated.

$$\text{concentration} = \frac{\text{contaminant released (cfm)}}{\text{dilution air required (cfm)}}$$

$$\frac{4.5 \text{ ft}^3}{\text{pint}} \times \frac{1 \text{ pint}}{4 \text{ minutes}} = \frac{1.13 \text{ ft}^3 \text{ released}}{\text{minute}}$$

$$\text{dilution air required} = \frac{\text{contaminant released (cfm)}}{\text{concentration}} = \frac{1.13 \text{ cfm}}{\frac{200}{10^6}} = 5650 \text{ cfm}$$

Answer: 5650 cubic feet of air/min — based on the oven's inlet, not the oven's outlet air temperature. (See problem 188.)

183. What is the maximum ground level concentration of gas from a stack designed for emergency venting of chlorine gas? The stack is 110 ft high. The maximum emission rate is five cubic feet of Cl_2 per second. The effective stack height is 120 ft due to the gas injection velocity into the ambient air. Assume that the wind velocity at the top of the stack is 10 mph.

The Bosanquet-Pearson equation is one which can be used to estimate maximum ground level concentration of a vent or stack-emitted air pollutant:

$$C_{max} = \frac{2.15 \times Q \times 10^5}{V \times H^2} \times \frac{p}{q}, \text{ where:}$$

p/q = x, y, and z plane diffusion parameters, often collectively taken as 0.63

H   = effective stack height = the sum of the physical stack height plus plume height resulting from the discharge velocity and the thermal buoyancy of the gas, in feet

C_{max} = maximum ground level concentration of the air pollutant, in ppm

Q   = emission rate of gas contaminant at ambient temperatures, in ft³/second

V   = wind velocity, in ft/sec

$$\frac{10 \text{ miles}}{\text{hour}} \times \frac{5280 \text{ feet}}{\text{mile}} \times \frac{\text{hour}}{60 \text{ minutes}} \times \frac{\text{minute}}{60 \text{ seconds}} = \frac{14.7 \text{ feet}}{\text{second}}$$

$$C_{max} = \frac{2.15 \times \frac{5\ ft^3}{second} \times 10^5}{\frac{14.7\ feet}{second} \times (120\ feet)^2} \times 0.63 = 3.2\ ppm\ Cl_2$$

Answer: 3.2 ppm chlorine gas at ground level. The approximate distance from the stack to the C_{max} point is 10 H or, in this case, $10 \times 120\ ft = 1200\ ft^2$. Check the residential set-back distances.

184. 97 air sample results obtained at an industrial process were linearly distributed when plotted on log-normal graph paper. The graph gave a 50% concentration as 1.03 ppm and an 84% concentration as 2.41 ppm. What is the geometric standard deviation for these air sample results?

$$GSD = \frac{84\%\ value}{50\%\ value} = \frac{2.41\ ppm}{1.03\ ppm} = 2.34$$

Answer: Geometric standard deviation = 2.34.

185. Air containing hydrogen sulfide gas was bubbled at one L/min for 12 min through an impinger containing 10 mL of 0.001 N iodine solution before the iodine was reduced. What was the H_2S concentration?

$S^{-2} + I_2 \rightarrow S^\circ + 2\ I^{-1}$

(mL) (N) = milliequivalents   (Iodine is the oxidant. H_2S is the reducing agent.)

(10 mL) (0.001 N) = 0.01 milliequivalent of I_2

1 meq I_2 is equivalent to 0.5 meq of H_2S

0.01 meq of I_2 is equivalent to 0.005 meq of H_2S

1 meq H_2S is equivalent to 24.45 mL H_2S at 25°C and 760 mm Hg

0.005 meq $H_2S \approx 0.122$ mL H_2S

$$\frac{0.122\ mL\ H_2S}{12\ L} = \frac{0.010\ mL}{L}$$

$$\frac{10\ mL}{1000\ L} = 10\ ppm\ H_2S$$

Answer: 10 ppm hydrogen sulfide gas (H_2S, a potent chemical asphyxiant).

186. A direct-reading air sampling instrument was calibrated with the following results:

calibration (true) concentration, ppm	instrument reading, ppm
10	7
30	63
70	91
200	320
500	570

What is the group correlation between the dependent and independent variables?

This can be computed by most standard hand-held scientific calculators or quite easily using computer spreadsheet analysis tools.

Answer: Very high, at r = 0.986. Although the group correlation is excellent, some individual readings are variable. Perhaps this instrument should be repaired or returned to the manufacturer for servicing and calibration.

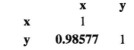

	x	y
x	1	
y	0.98577	1

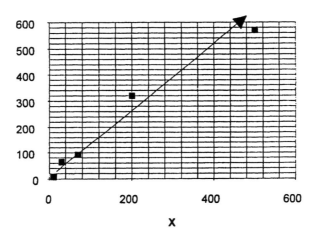

187. 1.3 pints of cyclohexane are uniformly evaporated from a large tray over a steam bath in an exhaust hood every seven min. The exhaust hood face dimensions are 2.5 × 4.5 ft. The average hood face velocity is 150 ft/min. The density of liquid cyclohexane is 0.78 g/mL. What is the cyclohexane vapor concentration in the exhaust air of the hood?

1.3 pints/7 min = 0.186 pint of cyclohexane evaporated per minute

$Q = A \times V = (2.5 \times 4.5 \text{ ft}) \times 150 \text{ fpm} = 1688 \text{ cfm}$

molecular weight $C_6H_{12} = 84$

$$\frac{1688 \text{ ft}^3/\text{minute}}{35.3 \text{ ft}^3/\text{minute}} = \frac{47.8 \text{ M}^3}{\text{minute}} \qquad \frac{68,623 \text{ mg}/\text{minute}}{47.8 \text{ M}^3/\text{minute}} = \frac{1435 \text{ mg}}{\text{M}^3}$$

$$\text{ppm} = \frac{\dfrac{\text{mg}}{\text{M}^3} \times 24.45}{\text{molecular weight}} = \frac{\dfrac{1435 \text{ mg}}{\text{M}^3} \times 24.45}{84} = 418 \text{ ppm cyclohexane}$$

Answer: 418 ppm of cyclohexane vapor are in the exhaust air.

188. Calculate the air volume in problem 182 if the discharge air temperature from the oven is 300°F.

Use the ratio of absolute temperatures, i.e., since the air expands with heat (but the mass remains the same), multiply the ratio of temperatures by a factor which is greater than one.

$$Q = 5650 \text{ cfm}_{\text{stp}} \times \frac{460°\text{F}+300°\text{F}}{460°\text{F}+70°\text{F}} = 8102 \text{ cfm}$$

Answer: 8102 ft³ of air at 300°F are discharged from the oven per minute.

189. Which use of organic vapor monitors best characterizes a production worker's 8-hr time-weighted average exposures to toluene and methyl isobutyl ketone (MIBK) vapors?

    a. Use one OVM for the entire exposure period.
    b. Use four OVMs sequentially for two hrs each.
    c. Use two OVMs sequentially for three to four hrs each.
    d. Use one OVM full shift on Tuesday. Repeat air sampling on Thursday.
    e. Use two OVMs at the start of work. Remove one at lunch time and the other at the end of work. Repeat this process on another day of exposure.
    f. Sample randomly with one OVM for any three-hr exposure period in the week. Repeat this on another day.

Answer: e.

190. An electrician who infrequently works in a TDI polyurethane foam production area develops asthma and no longer can work there without experiencing dyspnea, chest tightness, and an anxious feeling. Seventeen air samples obtained in the area over the past three years were between 0.001 and 0.014 ppm as eight-hr TWAEs. The TLV for TDI is 0.005 ppm with a ceiling concentration of 0.02 ppm. What appears most plausible?

    a. Since most of his exposures were well below the TLV, his illness must be due to nonoccupational factors (perhaps hobbies, avocations, home environment).
    b. He did not regularly work in the area, so it is unlikely he is sensitized to TDI.

c. The TDI sample test results cannot be used as proof or disproof of an existing illness. It appears he might be sensitized to TDI vapors. He must no longer work there. To do so might risk his health and, indeed, his life.

d. He is sensitized to the pyrolysis products released from hot insulation coating from smoldering electrical wires.

e. There is insufficient information to reach any tentative conclusions. It is not known if the air samples were eight-hr averages or short term samples. Obtain more data.

f. There is no work-relatedness of his condition to his exposures because no epidemiological studies have shown electricians to be at an increased risk of isocyanate sensitization.

Answer: c. (Review the *Preface* to the ACGIH *Threshold Limit Values* book.)

191. How many Pitot traverse readings should be taken when one measures velocity pressures in 6 to 40-in round stacks or ducts?

a. One carefully-placed center line reading × 0.8.

b. From five to 30 as long as 50% of measurements are within 10% of each other provided that the gas or air is less than 100°F.

c. Six.

d. It depends upon the water vapor content, density, temperature, and dust load of the air or gas being tested.

e. Two 10-point traverses with 20 sampling points and both at 90° to each other.

f. Two five-point traverses with 10 sampling points and both at 90° to each other provided that the gas or air is less than 100°F.

Answer: e. (Refer to the most recent edition of ACGIH's *Industrial Ventilation.*)

192. A degreasing tank loses trichloroethylene vapor to the general plant atmosphere because the tank does not have a mechanical local exhaust ventilation system. Based on the solvent purchase records, 1.1 gallons evaporate from this tank and the draining parts every eight hrs. Determine the emission rate/min to help design local exhaust ventilation for this 4 × 8-ft tank. The tank is covered to reduce evaporative losses except for 40 hrs during the week when the process is operating. The density of TCE = 1.46 g/mL.

$$\frac{1.1 \text{ gallon}}{8 \text{ hours}} \times \frac{3785 \text{ mL}}{\text{gallon}} \times \frac{1.46 \text{ g}}{\text{mL}} \times \frac{\text{hour}}{60 \text{ minutes}} \times \frac{1000 \text{ mg}}{\text{g}} = \frac{12{,}664 \text{ mg}}{\text{minute}}$$

Answer: The average TCE evaporation loss is 12,664 milligrams per minute. Refer to OSHA standards for open surface tanks (29 *CFR* 1910.94.).

193. A tracer gas was used to measure air infiltration rate into a building. The initial gas concentration was 0.01% (100 ppm). After one hr, it was 0.0012% (12

ppm). The dimensions of the building are $20 \times 40 \times 80$ ft. What was the infiltration rate?

$C = C_o\, e^{-(kt)/V}$, where:

C = tracer gas concentration after elapsed time, t
C_o = initial tracer gas concentration
e = 2.718
k = outside air infiltration rate
t = time
V = building volume
t = 1 hr = 60 min $V = 20 \times 40 \times 60$ ft = 64,000 ft^3

$$k = \frac{-\ln\left[\dfrac{C}{C_o}\right]}{\dfrac{t}{V}} = \frac{-\ln\left[\dfrac{0.0012}{0.01}\right]}{\dfrac{60\ \text{minutes}}{64,000\ \text{ft}^3}} = \frac{-\ln 0.12}{0.0009375} = \frac{-(-2.1203)}{0.0009375} = \frac{2263\ \text{ft}^3}{\text{minute}}$$

Answer: The average air infiltration rate was 2262 ft^3/min.

194. Which tracer gases are used in building air exchange measurements?

   a. sulfur dioxide, hydrogen, helium, nitrogen, methane
   b. helium, phosgene, argon, hydrogen selenide, nitrogen
   c. argon, nitrogen, helium, carbon dioxide, hydrogen
   d. methane, SF$_6$, ethane, nitrogen dioxide, helium
   e. sulfur hexafluoride, CO$_2$, helium, phosphine, arsine
   f. nitrogen, CO$_2$, SF$_6$, helium, argon

Answer: d. Helium and SF$_6$ gases are used most frequently.

195. An Orsat analysis of flue gas was 12.9% CO$_2$ in a combustion process using 40% excess air. Residual oxygen was 4.9% in this flue gas. What was the theoretical maximum concentration of CO$_2$ in the flue gas under these conditions?

$$\text{theoretical CO}_2 = \frac{\%\ \text{CO}_2\ \text{in flue gas sample}}{1-\left[\dfrac{\%\ \text{O}_2\ \text{in flue gas sample}}{21\%}\right]} = \frac{12.9\%\ \text{CO}_2}{1-\left[\dfrac{4.9\%\ \text{O}_2}{21\%\ \text{O}_2}\right]} =$$

$$\frac{12.9\%\ \text{CO}_2}{1-0.233} = 16.8\%\ \text{CO}_2$$

Answer: theoretical maximum concentration of CO$_2$ was 16.8%. Combustion efficiency, therefore, was (12.9%/16.8%) $\times$ 100 = 76.8%.

196. Isokinetic sampling for dust in a cement plant stack will be done with an air pump operating near one cfm. The velocity of the dust entering the nozzle must equal the velocity of the dust passing the nozzle to ensure representative dust sampling. The velocity of the stack gas is uniform and laminar at 37.65 ft/sec at the dust sampling point. What nozzle diameter is required?

$$\text{sampling rate} = 60 \times V x \frac{\pi \times d^2}{4 \times 144 \text{ in}^2/\text{ft}^2}, \text{ where:}$$

V = stack or duct velocity, feet per second

d = internal diameter of sampling probe, inches

Select a nominal meter stack sampling rate of 1 ft³/min.

1 cfm = 0.327 × (37.65 ft/sec) × d²

$$d^2 = \frac{\dfrac{1 \text{ ft}^3}{\text{minute}}}{(0.327)(37.65 \text{ feet/second})} = 0.0812 \text{ in}^2$$

$$d = \sqrt{0.0812 \text{ in}^2} = 0.285 \text{ inches}$$

The closest standard nozzle size is one with a 0.25 in. internal diameter. Therefore, calculation of meter rate for the smaller nozzle is necessary to give the same probe-to-stack velocity (and collection of a representative dust sample):

Q = V × A   (37.65 ft/sec) × (60 sec/min) = 2259 fpm

meter rate = 2259 fpm × 0.00545 × (0.25 in)² = 0.769 ft³/min

Answer: Air sampling pump meter rate = 0.77 ft³/min.

197. An Orsat gas analysis from burning natural gas is 10.4% CO_2, 2.9% O_2, and 86.7% N_2 by volume. The gas is 89% methane, 5% nitrogen, and 6% ethane by volume. What is a maximum theoretical percent CO_2 and percent excess air? Use 90% as the rough approximation of the ratio of dry products of combustion (ft³/ft³ of gas burned) to the air required for stoichiometric combustion (ft³/ft³ of gas burned).

$$\% \, CO_2 = \frac{10.4\% \, CO_2}{1 - \left[\dfrac{2.9\% \, O_2}{21\% \, O_2} \right]} = 12.06\% \, CO_2$$

$$\text{excess air} = \frac{12.06\% \, CO_2 - 10.4\% \, CO_2}{10.4\% \, CO_2} \times 0.90 = 14.37\% \text{ excess air}$$

Answers: 12.1% CO_2 and 14.4% excess air (oxygen). This calculation assists combustion engineers to achieve optimum oxidation of fuel and the production of minimum amounts of CO and other emissions in the flue and exhaust gases.

198. A pipe fitter working on a blast furnace inhaled 15,000 ppm carbon monoxide for one min (his first inhalation caused immediate collapse, and he continued to breathe the CO gas, although unconscious, for another 50 sec until he was rescued by a co-worker). The background ambient CO concentration throughout the steel mill was three ppm. What was his 8-hr TWAE assuming that he had no other exposures to CO that day? Where do you think his SCBA was?

$$1 \text{ min} \times 15,000 \text{ ppm CO} = 15,000 \text{ ppm-min}$$
$$479 \text{ min} \times 3 \text{ ppm CO} = \underline{1,437} \text{ ppm-min}$$
$$16,437 \text{ ppm-min}$$

16,437 ppm-min/480 min = 34 ppm

Answer: 34 ppm TWAE to CO, assuming he survived this massive, but brief, exposure. If he had not been rescued by a vigilant co-worker, should we have told his widow that his eight-hr TWAE complied with OSHA's 35 ppm PEL for CO?

199. Two adjacent gas lines rupture in a chemical plant exposing workers to an acrid aerosol. Five were hospitalized with severe pulmonary edema. A laboratory reconstruction of the accident revealed a maximum concentration of 6.3 milligrams of NH_4Cl/M^3 (STEL = 20 mg/M^3), 0.3 ppm Cl_2 (STEL = 1 ppm), and 3.6 ppm NH_3 (STEL = 35 ppm). The workers were not exposed for more than 10 min. What appears most plausible?

a. Since all exposures were well below PELs, these workers appear to have some other effects, perhaps psychosomatic "illness" or mass hysteria.

b. There were mistakes in the laboratory simulation tests.

c. These air contaminants cannot cause pulmonary edema.

d. The workers instead most probably have chemical pneumonitis.

e. The ammonia gas reacted with the chlorine gas to form chloramines — gases which are potentially far more injurious to the lungs than chlorine or ammonia. The laboratory tests failed to consider and detect chloramines.

f. These workers must have had pre-existing emphysema or bronchitis.

Answer: e.

200. The analytical detection limit for a hydrocarbon vapor with a molecular weight of 82 is five micrograms. The PEL is 150 ppm. What fraction of the PEL could be reported with a 100 mL air sample?

$$ppm = \frac{\dfrac{mcg}{L} \times 24.45}{molecular\ weight} = \frac{\dfrac{5\ mcg}{0.1\ L} \times 24.45}{82} = 14.9\ ppm$$

Answer: 14.9 ppm, or 10% of the PEL. Such a very small air sample might be obtained to measure peak exposures. 100 mL could be drawn through a small charcoal tube using a detector tube pump in two or three min. This method of air sampling could be used to augment full work shift air sampling using pumps on the worker's belts and BZ collection tubes.

201. The partial pressure of water vapor in an air sample at 22°C is 12.8 mm Hg. The saturation vapor pressure of water at this temperature is 19.8 mm Hg. What is the relative humidity of this air at 22°C and when heated to 28°C? The water vapor saturation pressure at 28°C is 28.3 mm Hg.

(12.8 mm Hg/19.8 mm Hg) × 100 = 64.6% R H at 22°C

(12.8 mm Hg/28.3 mm Hg) × 100 = 45.2% R H at 28°C

Answers: 64.6% and 45.2% relative humidities, respectively, at 22 and 28°C. If this air sample is cooled to 15°C, the air becomes saturated and the relative humidity is 100% because the saturation vapor pressure of the water is also 12.8 mm Hg.

202. A 5.3 cubic ft cylinder containing 2.7% arsine gas in nitrogen gas under high pressure bursts inside a 10 × 18 × 38-ft room with no ventilation. What is the gas concentration of AsH_3 in ppm after mixing? The molecular weight of $AsH_3 = 78$.

5.3 ft³ × 0.027 = 0.143 ft³ of arsine gas   10 × 18 × 38 ft = 6840 ft³

(0.143 ft³/6840 ft³) × 10^6 = 20.91 ppm

Answer: 21 ppm AsH_3. PEL = 0.05 ppm. The molecular weight is irrelevant.

203. A typical hydrocarbon emission rate for a stationary source coal combustion unit exceeding 10^8 BTU/hr capacity is 0.2 lb per ton of coal burned. What is the approximate annual hydrocarbon emission from a power plant burning 275 tons of coal per day in a steam generator with a 450 × 10^6 BTU/hr capacity?

$$\frac{275\ tons}{day} \times \frac{365.25\ days}{year} \times \frac{0.2\ lb\ HC}{ton} \times \frac{20,089\ lb\ HC}{year}$$

Answer: More than 10 tons of hydrocarbon vapors and gases are emitted per year.

204. The annual hydrocarbon emission for fluid catalytic units in petroleum refineries is typically 220 pounds for every 1000 barrels of fresh feed. What are the annual HC emissions for a unit with no CO boiler operating 250 days per year and consuming 12,000 barrels of fresh feed per day?

$$\frac{12,000 \text{ barrels}}{\text{day}} \times \frac{220 \text{ pounds}}{1000 \text{ bbl}} \times \frac{250 \text{ days}}{\text{year}} = \frac{660,000 \text{ lb HC}}{\text{year}}$$

Answer: 660,000 pounds (330 tons) of hydrocarbon vapor emissions per year.

205. A worker leans into a large gasoline storage tank that was recently filled to the 90% level. He collapses, but fortunately was seen by another worker and was revived by CPR. Analysis of the tank's head space shows 46.8 volume percent hydrocarbons. What was the most likely cause for his syncope and close brush with the "Grim Reaper?"

Answer: Since the tank's atmosphere contained 46.8% gasoline vapors, the balance was 53.2% air. 21% of this was oxygen ($0.21 \times 53.2\% = 11.2\%$ oxygen). The oxygen-deficient atmosphere plus the narcotic gasoline vapors most likely acted in a toxicologically-additive way to cause his collapse. OSHA regards an atmosphere containing less than 19.5% O_2 at sea level as oxygen-deficient.

206. A two-in wide by four-ft long slot hood exhausts air at a rate of 2000 ft/min. How many pounds of air are exhausted per hour by this hood? Assume that the hood is in an industrial plant in Tampa, FL.

$$\frac{2'' \times 48''}{144 \text{ in}^2/\text{ft}^2} \times \frac{2000 \text{ ft}}{\text{minute}} \times \frac{60 \text{ minutes}}{\text{hour}} \times \frac{0.075 \text{ lb}}{\text{ft}^3} \times \frac{6000 \text{ lb}}{\text{hour}}$$

Answer: 6000 pounds (three tons) of air are exhausted every hour.

207. The driver of a LPG fork lift truck had the following carbon monoxide exposures throughout her 10-hr work shift: 14 ppm for 3 3/4 hrs delivering parts to work stations, 16 ppm for 2 1/2 hrs returning empty pallets to the shipping and receiving dock, two ppm for two 15-min breaks, 135 ppm for 1 1/2 hrs inside railroad box cars, and 82 ppm for 1 3/4 hrs inside truck trailers. What was her TWAE to CO gas?

	ppm CO	×	time, hours	=	ppm-hours
delivering parts	14	×	3.75	=	52.5
returning pallets	16	×	2.5	=	40
coffee breaks	2	×	0.5	=	1
inside RR box cars	135	×	1.5	=	202.5
inside truck trailers	82	×	1.75	=	143.5
			10.0		439.5

439.5 ppm-hrs/10 hrs = 43.95 ppm TWAE

Answer: Her TWAE to CO is 44 ppm which exceeds the 35 ppm TLV. Try to reduce her exposure time inside trucks and rail cars.

208. The Agency for Toxic Substances and Disease Registry regards 0.26 microgram inorganic mercury per cubic meter of air as a concentration below which adverse health effects should not occur with continuous exposure. At a detection level of 0.1 microgram, how long must one sample at one L/min through an adsorbent tube to detect this concentration?

$$\frac{0.26 \text{ mcg}}{M^3} \times \frac{M^3}{1000 \text{ L}} \times \frac{1 \text{ L}}{\text{minute}} = \frac{0.00026 \text{ mcg}}{\text{minute}}$$

$$0.1 \text{ mcg} \times \frac{\text{minute}}{0.00026 \text{ mcg}} = 384.6 \text{ minutes}$$

Answer: Sample for 385 min assuming a 100% collection and desorption efficiency.

209. Air was sampled at an average flow rate of 2.34 L/min through a tared PVC membrane filter for three hrs and 27 min. The filter weight difference after air sampling was 2.32 milligrams. What was the airborne TSP concentration?

207 min × 2.34 L/m = 484.4 L

2320 mcg/484.4 L = 4.79 mg/M³

Answer: 4.79 milligrams of TSP/M³ of air.

210. A solvent blend is 20% ethyl acetate and 80% toluene (volume/volume). What is the vapor phase concentration of each component?

	ethyl acetate	toluene
density	0.90 g/mL	0.87 g/mL
vapor pressure	74 mm Hg	22 mm Hg
molecular weight	88.1	92.1

Use 100 mL of the solvent blend as a basis for calculations:

20 mL × 0.90 g/mL = 18 g ethyl acetate
80 mL × 0.87 g/mL = 69.6 g toluene

partial pressure of ethyl acetate =

$$\frac{\dfrac{18 \text{ g}}{88.1 \text{ g/mole}} \times 74 \text{ mm Hg}}{\dfrac{18 \text{ g}}{88.1 \text{ g/mole}} + \dfrac{69.6 \text{ g}}{92.1 \text{ g/mole}}} = \frac{15.119}{0.96} = 15.75 \text{ mm Hg}$$

partial pressure of toluene =

$$\frac{\dfrac{69.6\ g}{92.1\ g/mole} \times 22\ mm\ Hg}{\dfrac{18\ g}{88.1\ g/mole} + \dfrac{69.6\ g}{92.1\ g/mole}} = \frac{16.63}{0.96} = 17.32\ mm\ Hg$$

15.75 mm Hg + 17.32 mm Hg = 33.07 mm Hg total vapor pressure

(15.75 mm Hg/33.07 mm Hg) × 100 = 47.6% ethyl acetate vapor

(17.32 mm Hg/33.07 mm Hg) × 100 = 52.4% toluene vapor

Answers: 47.6% ethyl acetate vapor and 52.4% toluene vapor. Calculations were made using Raoult's law. Note how the minor (20%), but more volatile, component in the liquid phase increases and enriches in percent composition in the vapor phase (47.6%). Refer to problem 18 for discussion of application and deviations from Raoult's law.

211. A process evaporates 1.3 pints of benzene into the air of an empty room every eight hrs. The room is 10 × 16 × 25 ft. The density of benzene is 0.88 g/mL. The molecular weight of benzene is 78. What volume of dilution air is required to maintain the benzene vapor concentration below the NIOSH-recommended TWAE limit of 0.1 ppm?

Answer: Benzene is too hazardous and potentially toxic to rely on ventilation by dilution air as a control method. Reduction in the evaporation rate and the use of mechanical local exhaust ventilation are recommended to protect the health of exposed workers. Substitute a less hazardous solvent.

212. A gasoline powered electrical generator engine produces 47 cfm of total exhaust gases which contain 1.2% (vol/vol) of carbon monoxide into a 12-ft (h) × 30-ft (w) × 40-ft (l) garage. Ventilation for this garage is 0.3 cfm/ft² of floor area. Assuming a negligible concentration of CO in the make up air, what is the concentration of CO gas in the garage after 30 min of engine operation?

$$ppm\ CO = \frac{G \times 10^6 \left(1 - e^{-\left[\frac{Q}{V}\right]t}\right)}{Q}, \text{ where:}$$

G = rate of CO generation
V = volume of garage
Q = ventilation rate
t = time
e = 2.7813

47 cfm × 0.012 = 0.56 ft³ of CO generated/min (G)

0.3 cfm/ft^2 × 30 × 40 ft = 360 cfm (ventilation, Q)

V = 12 × 30 × 40 ft = 14,000 ft^3

$$\text{ppm CO} = \frac{0.56 \times 10^6 \left(1 - e^{-\left[\frac{360 \times 30}{14,400}\right]}\right)}{360} = 824 \text{ ppm CO}$$

Answer: 824 ppm carbon monoxide gas after 30 min. Ventilate with fresh air.

213. In the preceding example, if the generator is turned off after an industrial hygienist takes a CO detector tube air sample, how long will it take for the concentration to reduce to 20 ppm CO? Assume a steady, uniform ventilation of the garage.

$$\text{dilution ventilation time} = \frac{V}{Q} \times \left[\ln \frac{C_2}{C_1}\right] = \frac{14,400 \text{ ft}^3}{360 \frac{\text{ft}^3}{\text{minute}}} \times \left[\ln \frac{824 \text{ ppm}}{20 \text{ ppm}}\right] =$$

40 × ln 41.2 = 40 × 3.72 = 148.8 minutes

Answer: 149 min ≅ 2.5 hrs.

214. A worker is exposed to 87 ppm MEK and 23 ppm toluene vapor. Both are narcotic solvents with additive toxic effects. Their respective OSHA PELs are 200 and 100 ppm. What is the worker's additive exposure to these solvent vapors?

87 ppm + 23 ppm = 110 ppm

$$\frac{87 \text{ ppm}}{200 \text{ ppm}} + \frac{23 \text{ ppm}}{100 \text{ ppm}} = 0.44 + 0.23 = 0.67$$

Answers: 110 ppm total solvent vapors. Exposure is below the PEL for the mixture, however it exceeds the action level (67% of PEL for the mixture of solvent vapors; the action level is 50%). Industrial hygiene controls are warranted.

215. Several workers in a Wisconsin office building complain of eye irritation, sore throat, headache, and sinus "problems" every winter. Several air samples reveal concentrations of mineral spirits' vapor from an adjacent printing operation ranging from 0.3 to 1.5 ppm (geometric mean = 0.5 ppm). There are no other apparently remarkable industrial hygiene issues. What would you do?

a. Take more air samples.
b. Disregard their symptoms because vapor concentrations average less than 1% of the PEL.
c. Measure the relative humidity and consider the additive effects of dry air plus solvent vapors on their health and comfort. Suggest that they be seen by a physician. Consult with the physician after the examinations.

    d. Look for other air contaminants because mineral spirits' vapors cannot be the root cause of their mild complaints.

    e. Increase the air velocity flow through the work areas.

    f. Consider rotating some workers to other areas.

    g. Test the mineral spirits for benzene contamination.

    h. Their complaints are not statistically significant until 37% of the work force reports similar problems.

Answer: c.

216. A reactive gas is present in a plant atmosphere. This gas dissociates into a less hazardous gas at a predictable rate which is independent of the concentration that is present in the air. The gas also is removed by the plant ventilation system. A direct-reading gas measurement instrument reveals 25 ppm. Thirty min later, the gas concentration is 10 ppm. What is the combined decay constant of this gas with this ventilation?

$$\ln\left[\frac{\text{final concentration}}{\text{initial concentration}}\right] = -T_{1/2} \times \text{time}$$

$$\ln\left[\frac{10 \text{ ppm}}{25 \text{ ppm}}\right] = -T_{1/2} \times 30 \text{ minutes} \quad \ln 0.4 = -T_{1/2} \times 30 \text{ minutes}$$

$$\frac{-0.9163}{30 \text{ minutes}} = -T_{1/2} \quad\quad -T_{1/2} = -0.0305/\text{minute}$$

Answer: Combined decay constant = 0.0305 per min (i.e., 3.05%/min).

217. In the previous problem, the physical half-life of this gas is 40 min. Calculate the combined effective half-life (T_{eff}) and dilution ventilation half-life (T_v) of the gas.

$$\ln(1.2) = \frac{-0.0305}{\text{minute}} \times T_{eff} \quad\quad T_{eff} = \frac{-0.693}{-0.0305/\text{minute}} = 22.7 \text{ minutes}$$

$$T_{eff} = \frac{T_p \times T_v}{T_p + T_v}, \text{ where } T_p = \text{physical half-life}$$

$$22.7 \text{ minutes} = \frac{40 \text{ minutes} \times T_v}{40 \text{ minutes} + T_v}$$

$$908 + 22.7 \times T_v = 40 \times T_v \quad\quad 908/40 = (22.7 \times T_v/40) = T_v$$

$$22.7 + 0.568 \times T_v = T_v 22.7 = T_v (1 - 0.568 \quad\quad T_v = 52.5 \text{ min}$$

Answers: The combined effective half-life = 22.7 min. Ventilation half-life = 52.5 min.

218. An industrial hygiene chemist wants to prepare a 25-liter gas sampling bag with 35 ppm CO in it to calibrate an instrument. What volume of 1760 ppm CO gas must he dilute with CO-free air to obtain a 35 ppm concentration?

$$C_1 \times V_1 = C_2 \times V_2 \qquad V_1 = \frac{C_2 \times V_2}{C_1} = \frac{35\ \text{ppm} \times 25\ \text{L}}{1760\ \text{ppm}} = 0.497\ \text{L}$$

Answer: Dilute 497 mL of 1760 ppm CO to 25 L with CO-free air.

219. A room with dimensions of $12 \times 20 \times 26$ ft contains an ammonia compressor which leaks at a rate of 0.3 cfm. If 260 cfm is the volume of dilution air supplied to the room, by how much must the air supply be increased to ensure that the ammonia concentration does not exceed 10 ppm.

$$12 \times 20 \times 26\ \text{ft} = 6240\ \text{ft}^3 \qquad Q_o = 260\ \text{cfm} \qquad C = 10\ \text{ppm}$$

$$C_o = \frac{0.3\ \text{ft}^3}{6240\ \text{ft}^3} \times 10^6 = 48\ \text{ppm NH}_3 \qquad Q = 260\ \text{cfm} \times \frac{48\ \text{ppm}}{10\ \text{ppm}} = 1248\ \text{cfm}$$

Answer: Increase the dilution air by 988 cfm to 1248 cfm.

220. According to olfactory perceptions by room occupants, the odor intensity in a work area has "doubled." Assuming the source of the odor has a constant generation rate, by how much was the dilution ventilation apparently decreased? The dilution ventilation for the initial odor concentration was 16,400 cfm.

the equation, with I = the odor intensity, is:

$$\frac{I_1}{I_2} = \frac{\log \text{cfm}_2}{\log \text{cfm}_1}$$

This equation is similar to the Weber-Fechner law of physiological reactions in general, which is: a sensation is proportional to the logarithm of the stimulus.

$$\frac{1}{2} = \frac{\log \text{cfm}_2}{\log 16,400\ \text{cfm}} \qquad 0.5 = \frac{\log \text{cfm}_2}{4.215} \qquad \log \text{cfm}_2 = 2.108 \qquad \text{cfm}_2 = 128$$

Answer: Dilution air was reduced approximately from 16,400 cfm to 128 cfm. Or, stated another way, relatively tremendous volumes of clean air are required to dilute an odor to 50% of its perceived olfactory intensity. See problem 221.

221. In the preceding problem where the sensory odor intensity doubled with reduction of the dilution air from 16,400 cfm to 128 cfm, by how much did the actual odorant concentration apparently increase?

    16,400 cfm/128 cfm = 128 times

    Answer: 128 times. As the odor concentration increases to the second power, the apparent intensity doubles. Odor intensity varies with logarithm of concentration. Another psychophysical reaction covering the senses (olfaction, hearing, vision, taste, touch, etc.) is Stevens' law:

    $$I = k \ C^{\alpha}$$

    where I = perceived intensity of the sensation, or odor; k = a constant; and C = the physical intensity of the stimulus, the odorant. As a rule, sensation varies as a power function of the stimulus. In olfaction, α is less than one. Stevens' law has generally replaced the Weber-Fechner logarithmic law. (please refer to problem 220). Since α is invariably less than one, a change in concentration results in a smaller change in odor perception. Typically, α ranges from 0.2 to 0.7. With the exponent being 0.7, the odorant concentration must be reduced ten-fold to reduce the odor perception by a factor of 5. When α equals 0.2, there must be a three-thousand fold reduction in the concentration to achieve a five-fold reduction in odor perception.

222. An instrument calibrated at normal temperature and pressure (NTP = 25°C and 760 mm Hg) indicated 57 ppm at 16°C and 630 mm Hg pressure. What is the corrected concentration and the correction factor to be applied to all readings?

    ppm at NTP = ppm meter reading ×

    $$\frac{P}{760 \text{ mm Hg}} \times \frac{298 \text{ K}}{T} = 57 \text{ ppm} \times \frac{630 \text{ mm Hg}}{760 \text{ mm Hg}} \times \frac{298 \text{ K}}{289 \text{ K}} = 48.7 \text{ ppm}$$

    Answer: 49 ppm. The correction factor = (49 ppm/57 ppm) = 0.86.

223. Pot room air containing caustic soda dust (NaOH) was bubbled through a midget impinger at 0.92 L/m for 17 1/2 min. There were 13.2 mL of dilute H_2SO_4 acid solution in the impinger. Each milliliter of the acid will neutralize 0.43 micrograms of NaOH as indicated by a color change from purple to green. Assuming 100% collection efficiency by the impinger, what was the average airborne NaOH dust concentration when the indicator color (methyl purple) changed?

    0.92 L/m × 17.5 min = 16.1 L of air were sampled

    13.2 mL × 0.43 mcg/mL = 5.68 mcg NaOH

    5.68 mcg NaOH/16.1 L = 0.35 mcg/L

    Answer: 0.35 mg NaOH dust per cubic meter of pot room air.

224. Sulfur hexafluoride, SF_6, was used to measure reentry of the exhaust air from a hood into a plant. The hood had an exhaust ventilation rate of 18,500 cfm. SF_6 gas was released at a rate of 0.05 cfm into the hood's exhaust system. The plant had a general replacement air flow rate of 77,600 cfm. What was the percent reentry if the SF_6 steady-state concentration in the plant was 0.004 ppm?

$$\text{exhaust concentration} = \frac{0.05\,\text{cfm}}{18,500\,\text{cfm}} \times 10^6 = 2.7\,\text{ppm}$$

$$\text{ventilation dilution factor} = \frac{77,600\,\text{cfm}}{18,500\,\text{cfm}} = 4.19$$

The concentration of SF_6 measured inside the plant was 0.004 ppm. Contaminant dilution, therefore, is 2.7 ppm/0.004 ppm = 675 times.

Re-entry of contaminants from the hood into the plant, given as a fraction of the released SF_6, = 4.19/675 = 0.0062.

Answer: 0.0062 × the released amount, or 0.62% reentry.

225. Contrast the ratio of surface areas of one micron diameter particles to particles with a diameter of three microns.

surface area of a sphere = $\pi \times$ (diameter)^2

for a one micron Ø particle: $\pi \times (1\,\mu)^2 = 3.1416$ square microns

for a three micron Ø particle: $\pi \times (3\,\mu)^2 = 28.274$ square microns

28.274 microns^2/3.1416 microns^2 = nine (the square of the diameter, e.g., for a five micron diameter particle, the surface area would be 25 times greater).

Answer: Nine times greater. As particle diameter increases three-fold, the surface area increases six-fold. If the diameter increases, e.g., six-fold, the surface area increases $(6)^2$-fold, or 36 times.

226. A virulent strain of infectious virus has a molecular weight of 55×10^6. This virus, resistant to desiccation, is in an aerosol containing 12 picograms of the virus per cubic meter of air. How many virus molecules are inhaled with every liter of viral aerosol?

$$\frac{12\,\text{picograms}}{M^3} = \frac{0.000012\,\text{microgram}}{1000\,L} = \frac{0.000000012\,\text{microgram of virus}}{L}$$

$$\frac{1.2 \times 10^{-14}\,g}{L} \times \frac{\text{mole}}{55 \times 10^6\,\text{gram}} \times \frac{6.023 \times 10^{23}\,\text{molecules}}{\text{mole}} = \frac{1.3 \times 10^2\,\text{molecules}}{L}$$

Answer: 130 molecules of virus per liter of inhaled air.

227. A worker sprayed banana plants in Ecuador with a pesticide. A breathing zone air sample using a tared PVC membrane filter was used to collect the spray aerosol mist. Assume 35% of the collected particulate was not pesticide and 7% of the pesticide evaporated from the filter during air sampling. If the air sampling rate was 2.1 L/min for seven hrs and 41 min, what was the sprayer's pesticide exposure if the initial filter weight was 45.46 mg and the final filter weight was 49.13 mg?

(2.1 L/m) × 461 min = 968.1 L

49.13 mg – 45.46 mg = 3.67 mg total aerosol particulate on the PVC filter

3.67 mg × 0.65 = 2.39 mg pesticide/M^3

(2.39 mg pesticide/M^3) × 1.07 evaporation factor = 2.56 mg/M^3

Answer: 2.56 milligrams of pesticide per cubic meter of breathing zone air.

228. During a laboratory instrument calibration for airborne oxidant (I_2 vapor), 45.67 mg of iodine crystals were placed in a 0.17 ft³/min air stream. The crystals were removed after 176 min and found to weigh 43.79 mg. What was the average concentration of sublimed iodine vapor in the air stream in ppm? The molecular weight of I_2 = 253.8.

$$\frac{0.17\ ft^3}{minute} \times \frac{28.3\ L}{ft^3} \times 176\ minutes = 846.7\ L \qquad 45.67\ mg - 43.79\ mg = 1880\ mcg$$

$$\frac{1880\ mcg}{846.7\ L} = \frac{2.22\ mcg}{L} \qquad ppm = \frac{\dfrac{2.22\ mcg}{L} \times 24.45}{253.8\ grams/mole} = 0.21\ ppm$$

Answer: 0.21 ppm sublimed I_2 vapor in the effluent air stream.

229. A hospital requires humidification for occupant comfort. Engineering calculations are based on an outdoor temperature of 41°F and a relative humidity of 20%. An indoor temperature at a relative humidity of 50% is selected as a design objective. Additional increases in humidity and dry bulb temperature will be gained from the occupants, other water vapor sources, and process lighting and heating, etc. How much H_2O vapor must be evaporated into the influent outside air to achieve this humidity? Water vapor saturation is 6.77 g/M^3 at 41°F and 17.26 g/M^3 at 68°F. This hospital is in New York City.

density of water vapor in 41°F air = 0.20 × 6.77 mg/M^3 = 1.354 g/M^3

density of water vapor in 68°F air = 0.50 × 17.26 g/M^3 = 8.63 g/M^3

41°F = 5°C, and 68°F = 20°C

One M^3 of air at 41°F expands to: $\dfrac{273\,K + 20}{273°\,K + 5} = \dfrac{293\,K}{278\,K} = 1.054\,M^3$ at 68°F.

Water vapor in 1.054 M^3 at 68°F = $1.054\,M^3 \times \dfrac{8.63\,g}{M^3} = 9.10$ grams.

Amount of water vapor to be added to each cubic meter of air at 41°F = 9.10 g − 1.35 g = 7.75 g of H_2O.

Answer: 7.75 g of water vapor must be added to every cubic meter of 41°F outdoor air at 20% relative humidity.

230. A pressure vessel contained carbon monoxide compressed to 17 atmospheres at 43°F. The tank ruptures at what was believed to be an internal CO temperature of 300°F. People die from blast effects and chemical asphyxiation. Accident reconstruction attempts to determine the probable internal pressure of the CO when the tank blew apart.

Since the tank volume is constant, $V_i = V_f$, then:

$$\frac{P_i V_i}{T_i} = \frac{P_f V_f}{T_f} \text{ becomes } \frac{P_i}{T_i} = \frac{P_f}{T_f} = \frac{17 \text{ atmospheres}}{(43 + 460)° \text{ R}} = \frac{P_f}{(300 + 460)° \text{ R}}.$$

Pf = 25.69 atmospheres

Answer: Tank rupture occurred near an internal pressure of 26 atmospheres. Correction for the nonideal gas behavior at this high pressure might be required

231. A mixture of gases at 25°C contains nitrogen at 160 mm Hg, methane at 212 mm Hg, hydrogen at 110 mm Hg, ethane at 210 mm Hg, and propane at 195 mm Hg. What is the total pressure of the gas mixture and the percent noncombustible gas in the mixture?

(160 + 212 + 110 + 210 = 195) mm Hg = 887 mm Hg

Nitrogen is the only noncombustible, nonflammable component:

$$\frac{100 \text{ mm Hg}}{887 \text{ mm Hg}} \times 100 = 18.04\% \text{ N}_2$$

Answer: 18% nitrogen. 82% flammable gases. 887 mm Hg total pressure.

232. A pressure vessel contains 3700 kilograms of ethane gas at one atmosphere and 10°C. How much ethane could it contain at 30°C and 14 atmospheres?

The tank volume is constant, therefore: $\dfrac{P_i}{T_i} \propto \dfrac{P_f}{T_f}$.

$$\dfrac{\dfrac{14\ \text{atm.}}{(273\ K+30°\ C)\ K}}{\dfrac{1\ \text{atm.}}{(273\ K+10°\ C)\ K}} = \dfrac{0.046205}{0.003534} = 13.074 \quad 13.074 \times 3700\ kg = 48,374\ kg\ C_2H_6$$

Answer: 48,374 kilograms of ethane.

233. A boll weevil returns to the nest from the fields with 600 nanograms of pesticide with a molecular weight of 378 on each foot. He dies from acute cholinesterase poisoning. The subterranean nest — which has no ventilation — is 11 cubic centimeters. What is the concentration of the pesticide vapor in their nest air after evaporating from his feet? If the STEL for adult weevils is one ppm (15 min), are other weevils in the same nest in danger?

one nanogram = 10^{-9} g = 10^{-3} micrograms

(600 nanograms/ft) × 6 ft = 3600 nanograms

3.6 micrograms of pesticide evaporate from his feet while his wife mourns and prepares to bury his contaminated remains and hard-hat in a hazardous waste landfill.

$$11\ cm^3 = 11\ mL \times \dfrac{L}{1000\ mL} = 0.011\ L \qquad \dfrac{3.6\ mcg}{0.011\ L} = \dfrac{327.3\ mcg}{L}$$

$$ppm = \dfrac{\dfrac{327.3\ mcg}{L} \times 24.45}{378} = 21.2\ ppm$$

Answer: 21.2 ppm. The other weevils are in danger as they have continuous exposure.

234. OSHA's action level for asbestos fibers is 0.1 fiber per cc as an eight-hr time-weighted average exposure. Assuming a worker inhales 10 cubic meters of air containing 0.1 f/cc during his work shift, how many asbestos fibers will he inhale?

$$10\ M^3 \times \dfrac{1000\ L}{M^3} \times \dfrac{1000\ cm^3}{L} \times \dfrac{0.1\ fiber}{cm^3} = 10^6\ fibers$$

Answer: At least 1,000,000 asbestos fibers are inhaled during the 8-hr work shift. This only includes those fibers with a length-to-width aspect ratio of 3-to-1 and with a length longer than five microns. If the fibers having the same aspect ratio but shorter than five microns are included, the total number of asbestos fibers would substantially increase.

235. 30% of the workers in a metal machining plant simultaneously develop an acute illness and are too ill to work Their symptoms include fever, myalgia, headache, dyspnea, and extreme fatigue. Inhalation of aerosols from the metal working fluid is believed to be the source of an apparent epidemic bacterial or viral infection. Analysis of the coolant reveals the presence of 10^8 aerobic bacteria per mL and 130,000 *Legionella pneumophillae* per mL. Workers were exposed to a geometric mean concentration of 5.8 mg total airborne mist and aerosol/M^3. If the coolant metal working fluid was 95% water and had a density of one g/mL, how many *L. pneumophillae* bacteria were inhaled in each liter of air at the geometric mean concentration?

Consider the number of bacteria in one liter of air.   5.8 mg/M^3 = 5.8 mcg/L

The volume of coolant mist generated and required to be filtered by the sampling device, V, to collect 5.8 mcg of particulates (including all bacteria) is:

$V \times (1.00 - 0.95) = 5.8$ mcg

$$V = \frac{5.8 \text{ mcg}}{0.05} = 116 \text{ mcg of total mist (includes the aqueous portion)}$$

$$116 \text{ mcg} \times \frac{g}{10^6 \text{ mcg}} \times \frac{1.0 \text{ mL}}{1.0 \text{ g}} = 0.000116 \text{ mL}$$

$$0.000116 \text{ mL} \times \frac{130,000 \text{ L. pneumophillae}}{\text{mL}} = 15.08 \text{ liters}$$

Answer: 15 *L. pneumophillae* bacteria inhaled per liter of air. It has been estimated that as little as one *L. pneumophillae* bacterium in 50 L of inhaled air can be an infectious dose in a susceptible human host.

236. Derive an equation to calculate the half-life for an air contaminant in a room with a volume (V), uniform ventilation (Q), original concentration (C_o), and concentration at half-life time (C) after generation stops (G = 0). t = the time required to reach the half-life concentration.

The volume of the space and the ventilation rate determine the half-life.

$$C = C_o \times \left(e^{-\left[\frac{Qt}{V}\right]} \right) \qquad \text{at one half-life: } C = 0.5 \, C_o$$

$$\frac{0.5 \, C_o}{C_o} = e^{-\left[\frac{Qt}{V}\right]} \qquad t = -\left[\frac{V}{Q}\right] \ln 0.5 = 0.693 \left[\frac{V}{Q}\right]$$

Answer: The air contaminant half-life $= 0.693 \left[\dfrac{V}{Q}\right]$.

237. Evaluate a plant that is 20-ft (h) × 45-ft (w) × 60-ft (l) which has a general ventilation rate of 1.3 cfm per square foot of floor area. The ventilation is evenly distributed. If the CO concentration in the plant is 200 ppm, what is the CO concentration after three half-lives? What is the half-life for this plant under these conditions?

$$\frac{1.3 \text{ ft}^3}{\frac{\text{minute}}{\text{ft}^2}} \times (45 \times 60 \text{ ft}) = \frac{3510 \text{ ft}^3}{\text{minute}} \quad \text{plant volume} = 20 \times 45 \times 60 \text{ ft} = 54,000 \text{ ft}$$

$$\text{half-life} = 0.693 \left[\frac{V}{Q}\right] = 0.693 \left[\frac{54,000 \text{ ft}^3}{3510 \text{ cfm}}\right] = 10.66 \text{ minutes}$$

three half-lives = (10.66 min/half-life) × 3 = 32 min

200 ppm × $(0.5)^3$ = 25 ppm

Answer: The concentration of CO gas remaining in the plant atmosphere after three half-lives (32 min) of ventilation is 25 ppm assuming that CO release and generation stops when the ventilation begins. The half-life is 10.7 min.

238. A dry cleaner operator was exposed to 11 ppm "perc" vapor for 3.5 hrs at a spotting bench, 15 ppm for 1.5 hrs while sorting and loading clothes, 49 ppm for 2.5 hrs while removing and hanging clothes, and two ppm for 2.5 hrs taking orders and returning cleaned clothing. What was her TWAE to perchloroethylene vapor?

job	C	×	T	=	CT
spotting	11 ppm	×	3.5 hrs	=	38.5 ppm-hrs
sorting and loading	15 ppm	×	1.5 hrs	=	22.5 ppm-hrs
removing and hanging	49 ppm	×	2.5 hrs	=	122.5 ppm-hrs
taking orders	2 ppm	×	2.5 hrs	=	5.0 ppm-hrs
			10.0 hrs		188.5 ppm-hrs

TWAE = 188.5 ppm-hrs/10 hrs = 18.85 ppm "perc" using Haber's law

Answer: Her 10-hr TWAE to perchloroethylene vapor was 19 ppm.

239. A silver membrane air filter taken from a topside coke oven worker was extracted with benzene. This tared filter weighed 78.57 mg before sampling and 82.97 mg after air sampling. The same filter weighed 80.76 mg after extracting with warm benzene. If air was sampled at 2.07 L/min for 463 min, what was the worker's exposure to total airborne particulates and to the benzene-soluble fraction of the airborne coke oven emissions?

(2.07 L/m) × 463 min = 958.4 L   82.97 mg – 78.56 mg = 4.41 mg

$$\frac{4410 \text{ mcg total particulates}}{958.4 \text{ L}} = \frac{4.60 \text{ mg TSP}}{M^3}$$

$82.97 \text{ mg} - 80.76 \text{ mg} = 2.21 \text{ mg COE/M}^3$

Answers: 4.60 mg total suspended airborne particulates/M^3 and 2.31 mg coke oven emissions/M^3 (50% of total).

240. A balloon man sold 12 balloons filled with 20% ammonia and 80% helium instead of pure helium to youngsters at a birthday party. He instructed them to prick every balloon at the same time for a surprise. Each balloon was filled with six L of the mixed gas. What was the burst concentration of ammonia gas when the children pricked the balloons in a 8 × 16 × 18-ft room with no ventilation?

$$12 \text{ balloons} \times \frac{6 \text{ liters}}{\text{balloon}} \times 0.2 = 12.4 \text{ liters of ammonia}$$

$$8 \times 16 \times 18 \text{ ft} \times \frac{28.3 \text{ L}}{\text{ft}^3} = 65,249 \text{ liters} \qquad \frac{14.4 \text{ L}}{65,249 \text{ L}} \times 10^6 = 220.6 \text{ ppm NH}$$

Answer: The average concentration in the room would have been about 221 ppm. However, the burst concentration near each balloon was less than 200,000 ppm — the ammonia concentration in each balloon. This would cause coughing and tearing.

241. Which OSHA-approved gases and vapors are used to odorize LPG?

a. ethyl mercaptan, thiophane, amyl mercaptan

b. thiophene, butyl mercaptan, methyl mercaptan

c. ethane thiol, isoamyl mercaptan, mercaptobenzothiazole

d. SO_2, butyl mercaptan, acetylene

e. propane thiol, thiophene, thiophane, hydrogen sulfide

Answer: a.

242. Ethyl mercaptan is typically used as an odorant for natural gas and liquefied fuel gases at a concentration of about eight pounds per million cubic feet of gas. What is the concentration of C_2H_5-SH in natural gas in ppm? Density of ethyl mercaptan (ethanethiol) is 0.839 g/mL. The molecular weight of ethyl mercaptan is 62.13.

$$\frac{8 \text{ lb}}{10^6 \text{ ft}^3} \times \frac{454 \text{ g}}{\text{lb}} \times \frac{1000 \text{ mg}}{\text{g}} \times \frac{35.3 \text{ ft}^3}{M^3} = \frac{128 \text{ mg}}{M^3}$$

$$\frac{\dfrac{128 \text{ mg}}{M^3} \times 24.45}{62.13} = 50 \text{ ppm } C_2H_5SH$$

Answer: 50 ppm ethyl mercaptan. This greatly exceeds the odor threshold of approximately 0.5 ppb (0.0005 ppm) which most people can detect. The source concentration must be much greater than the odor thresholds because substantial dilution occurs between the leak and the person detecting the gas. An important consideration, of course, is the notion that "If I'm smelling this odorant at, say 1 ppb and the leak is 50 or more feet away, could the concentration of LPG exceed its LEL somewhere between the source and me?" Any ignition sources?

243. What is the saturation concentration of "Heptachlor" in air at NTP? The vapor pressure of this pesticide is 0.0003 mm Hg, and its molecular weight is 373.4. Does the saturation concentration exceed the OSHA PEL of 0.5 mg/M³ (Skin)?

$$\frac{0.0003 \text{ mm Hg}}{760 \text{ mm Hg}} \times 10^6 = 0.395 \text{ ppm}$$

$$\frac{\text{mg}}{M^3} = \frac{\text{ppm} \times \text{ molecular weight}}{24.45} = \frac{0.395 \times 373.4}{24.45} = \frac{6.03 \text{ mg}}{M^3}$$

Answer: Six mg/M³, or a vapor saturation concentration of "Heptachlor" about 12 times the OSHA PEL. However, in the field, the typical exposures to this pesticide are to mist — not solely to vapor — and evaluations must include an appraisal of mist, vapor, and skin exposures among other issues of industrial hygiene.

244. An explosion occurred in a tank where LPG was used as fuel for a heater to dry the tank's interior. A forensic accident investigation found the LPG cylinder outside of the tank and a hose that fed propane gas into the tank heater. The heater tipped over, the ignition flame was extinguished, and leaking propane gas accumulated inside the tank. From the tare weight and the water capacity stamped on the LPG container, reweighing indicated 2.8 gallons of LPG had vaporized inside the eight-ft diameter, 24-ft long cylindrical tank. The source of ignition appeared to be an ineffective exhaust fan located near the far end of the tank. The density of LPG is 0.51 g/mL. What could have been the maximum concentration of propane gas in the tank?

volume of cylinder =

$$\pi(r)^2 (h) = \pi(4 \text{ ft})^2 (24 \text{ ft}) = 1206.37 \text{ ft}^3 \times \frac{28.3 \text{ L}}{\text{ft}^3} = 34,164 \text{ liters}$$

$$2.8 \text{ gallons} \times \frac{3.785.3 \text{ mL}}{\text{gallon}} \times \frac{0.51 \text{ g}}{\text{mL}} = 5405.5 \text{ grams}$$

$$\frac{5405.5 \text{ g}}{34,164 \text{ L}} = \frac{0.1582 \text{ g}}{\text{L}} = \frac{158.2 \text{ mg}}{\text{L}} = \frac{158,200 \text{ mg}}{\text{M}^3}$$

molecular weight of $C_3H_8 = 44$ $ppm = \dfrac{\dfrac{158,200 \text{ mg}}{\text{M}^3} \times 24.45}{44} = 87,909$

Answer: 87,909 propane gas in the tank (= 8.79%). LEL = 2.2% = 22,000 ppm. UEL = 96,000 ppm. 8.79% approaches the "richer" end of the LEL-UEL range and, therefore — while certainly explosive — the magnitude of any explosion of gas or vapor increases as the concentration approaches the stoichiometric mid-range of explosibility.

245. The industrial hygienist often must explain how small a ppm is, how tiny are these dust particles, "What's a microgram?", how short a five micron fiber is, and "What does 0.002 mg Be/M^3 mean?", etc. One good approach that can be used is to relate the familiar to the unfamiliar. As an example, to share the notion of the OSHA PEL for lead of 50 micrograms per cubic meter, one could relate the weight of the artificial sweeteners (e.g., Equal®) found in the little paper packages we see along with packets of sugar in restaurants. This could be used to demonstrate the dust concentration inside a building the size of, say, a football field 300 ft by 160 ft by 14 ft high. The weight of artificial sweeteners in those little packets is one g. Assuming that the sweetener is ultrafine lead powder, what would be the airborne lead dust concentration if it was suspended in the air of this gridiron-sized building?

$$\frac{1 \text{ gram} \times \dfrac{10^6 \text{ micrograms}}{\text{gram}}}{\dfrac{300 \times 160 \times 14 \text{ ft}}{35.315 \text{ ft}^3}} = \frac{53 \text{ mcg Pb}}{\text{M}^3}$$
$$\frac{}{\text{M}^3}$$

Answer: 53 micrograms of lead per cubic meter, or slightly over the OSHA PEL of 50 mcg/M^3 and 23 mcg/M^3 above the "action level" for work place airborne lead. And, such a finely-divided dust at this concentration would be essentially invisible.

246. A 70 kilogram man swallowed a one g chunk of sodium cyanide which fell from an overhead beam into a sandwich he was eating. Immediately, the soluble cyanide salt converted into HCN gas when contacting his gastric HCl. This caused him to belch 400 mL of gas containing 2800 ppm HCN gas. How much cyanide was left in his body after belching? Assume an excess of hydrochloric acid in his stomach and that the belched gas was totally expelled from his body.

NaCl + HCl → HCN↑ + NaCl

molecular weights = 49 for NaCN and 27 for HCN

One g NaCN/(49 g/mol) = 0.02041 mol NaCN; therefore, 0.02041 mol of HCN was produced.

$$\frac{mg}{M^3} = \frac{ppm \times molecular\ weight}{24.45} = \frac{2800\ ppm \times 27}{24.45} = \frac{3092\ mg}{M^3} = \frac{3092\ mcg}{L}$$

$$\frac{3092\ mcg}{L} \times 0.4\ L = 1237\ mcg\ HCN$$

$$\%\ CN^- \ in\ HCN = \frac{26}{27} \times 100 = 96.3\%$$

$$\%\ CN^- \ in\ NaCN = \frac{26}{49} \times 100 = 53.1\%$$

1237 mcg HCN × 0.963 = 1191 mcg CN⁻ = 1.19 mg CN⁻

1000 mg NaCN × 0.531 = 531 mg CN⁻

531 mg CN⁻ − 1.19 mg CN⁻ = 529.8 mg CN⁻

Answer: 529.8 milligrams of cyanide remain as HCN in his body. This equals 529.8 mg/70 kg = 7.6 mg CN⁻/kg, or (7.6 mg/kg) × (1/0.531) = 14.3 mg NaCN/kg. A LDL_0 of 2.857 NaCN/kilogram has been reported for man. HURRY. HURRY. — quickly administer the antidotes amyl nitrite and sodium nitrite. Keep the person warm. Give CPR, oxygen, etc. Call a doctor, EMS, or local Poison Control Center. Do not become a victim yourself by inhaling exhaled HCN of the victim during mouth-to-mouth CPR.

247. An industrial hygienist evaluated a maintenance worker's exposures to methyl chloroform vapors during the removal of sludge from a solvent degreasing tank. A 45-min time-weighted average exposure was 350 ppm (1900 mg/M³, the eight-hr OSHA PEL). A few days later, the same worker was overcome from methyl chloroform at a spray degreasing operation and admitted to a hospital. Reconstruction of the exposure ironically showed that he was also exposed to 350 ppm for 45 min. Charcoal tubes and air sample pumps were used for both studies. What could account for the man's illness from one exposure and not the other apparently identical exposure?

Answer: The first exposure was solely to methyl chloroform vapors, while the second most likely included the inhalation of mist particles as well as vapors. There is no good air sampling technique which can distinguish between both MC vapors and MC mist particles in an aerosol mixture. Inhalation of respirable MC particles is expected to be far more irritating and injurious to lungs than vapor alone. Compare, for example, the eye irritation expected from vapor exposure to that resulting from direct contact of liquid solvent on the cornea. The glycolipid-covered membranes in the alveoli are more susceptible to injury by mist particles than from vapor molecules. In the evaluation of exposures to mixed aerosols, the industrial hygienist must be vigilant for variations in biological responses from what appear to be the "same" exposure concentrations. Observation of a mist cloud in breathing zones of workers is a clue that exposures are to a two-phase aerosol.

248. A calibrated rotameter indicated an air flow rate of 2.37 L/min at 7:07 am. This rotameter — used in a sampling system for airborne dust — indicated 2.18 L/m at 4:02 pm. How much air was sampled?

$$\frac{2.37\,L/min + 2.18\,L/min}{2} = 2.28\,L/min \text{ average air flow rate}$$

elapsed time = 15:62 − 7:07 = 8 hrs and 55 min = 535 min

(2.28 L/m) × 535 min = 1219.8 L

Answer: Approximately 1220 L of air were sampled (1.22 M³).

249. The PEL for a dust is 0.3 mg/M³ as total suspended particulates (TSP). The detection limit is five micrograms. At an air sampling rate of 1.06 L/m, how long must one sample to detect 10% of the PEL?

$$\text{minimum sampling time (in minutes)} = \frac{\text{analytical sensitivity (in mcg)}}{0.1 \times PEL\,(mcg/L) \times L/min} =$$

$$\frac{5 \text{ micrograms TSP}}{0.1 \times 0.3\,mcg/L \times 1.06\,L/min} = 157.2 \text{ minutes}$$

Answer: Sample at least two hrs and 38 min.

250. A crane used in construction of a skyscraper raised a compressed gas bottle containing 200 ft³ of 100% acetylene to welders on the 41st floor. The bottle burst in free space at the 20th floor. If the gas concentration was 100,000 ppm four ft from the cylinder microseconds after bursting, what concentration of gas could be expected eight, 12, and 16 ft away (i.e., spheres with 16, 24, and 32 ft diameters)? Assume no wind or thermal air currents will immediately dilute the gas. Assume uniform energy dispersion as the compressed cylinder of acetylene ruptured.

The inverse square law would appear to apply here; that is, the concentration will decrease as the inverse of the square of the distance from the source. This is an expanding sphere of acetylene gas decreasing in concentration and energy as the inverse of the square of the distance from the epicenter.

in the gas cylinder: 100% acetylene (i.e., 1,000,000 ppm)

at four ft: 100,000 ppm

at eight ft: $(1/2^2) \times 100{,}000$ ppm $= 25{,}000$ ppm (i.e., two $\times$ four ft)

at 12 ft: $(1/3^2) \times 100{,}000$ ppm $= 11{,}111$ ppm (three $\times$ four ft)

at 16 ft: $(1/4^2) \times 100{,}000$ ppm $= 6250$ ppm (four $\times$ four ft)

Answers: About 25,000 ppm, 11,000 ppm, and 6000 ppm of C_2H2 at 8, 12, and 16 ft, respectively.

251. Consider a one cubic centimeter solid crushed into one micrometer cubes. How many particles are created? How does the total surface area of the small particles compare to the surface area of the original cube? What are the implications for alveolar surface activity?

 original: 6 facets $\times$ (1 cm $\times$ 1 cm)/facet $= 6$ cm^2 1 cm $= 104$ microns/side

 after crushing to tiny particles: $10^4 \times 10^4 \times 10^4 = 10^{12}$ particles

 10^{12} particles $\times$ 6 microns^2/particle $= 6 \times 10^{12}$ microns^2 $= 6$ M^2 (and biological activity$\uparrow$)

 Answers: One large particle becomes 1,000,000,000,000 tiny particles, and the surface area increases from 6 cm^2 to 60,000 cm^2.

252. NIOSH reports that the inhalation of a single tubercle bacillus can cause an active tuberculosis lesion (*Occupational Respiratory Diseases*, 1986). This bacterium weighs as little as 10^{-13} g. A careless laboratory worker was exposed to an aerosol of tubercle bacilli estimated to be about 20,000 microbes per cubic meter. She inhaled at most 30 L of bio-contaminated air before leaving the facility. How many bacilli could she have inhaled? What was her total inhaled mass of this bacillus?

 20,000 microbes/M^3 = 20,000/1000L = 20 microbes/liter of air

 $\dfrac{20 \text{ microbes}}{\text{L}} \times 30$ L of inhaled air $= 600$ microbes dose

 600 microbes $\times$ 10^{-13} g/microbe $= 6 \times 10^{-11}$ g/microbe $= 60$ picograms

 Viruses, by comparison, weigh about one attogram $= 10^{-18}$ g $= 10^{-3}$ femtogram.

 Answer: 600 tubercle bacilli microbes. 60 picograms of tubercle bacilli.

253. Jerry returns home from the dry cleaners with 26 pounds of clothing and hangs it in a 3 $\times$ 4 $\times$ 8-ft closet with about 20% contents. One milliliter of perchloroethylene remains in the clothing after he removes the plastic garment bags and hangs the clothing. He shuts the door. The closet has no ventilation. Tom, the family cat, is inside the closet. What is Tom's exposure when all of the perchloroethylene evaporates? The molecular weight and "perc's" density = 165.8 and 1.62 g/mL, respectively.

$3 \times 4 \times 8 \text{ ft} = 96 \text{ ft}^3$ $96 \text{ ft}^3 - (0.20 \times 96 \text{ ft}^3) = 76.8 \text{ ft}^3$

$$\frac{1.62 \text{ g ''perc''}}{76.8 \text{ ft}^3} = \frac{1,620,000 \text{ mcg}}{76.8 \text{ ft}^3} \times \frac{\text{ft}^3}{28.32 \text{ L}} = \frac{744.8 \text{ mcg}}{\text{L}}$$

$$\frac{\dfrac{744.8 \text{ mcg}}{\text{L}} \times 24.45 \text{ L/g-mole}}{165.8 \text{ g/g-mole}} = 109.8 \text{ ppm ''perc'' vapor}$$

Answer: 110 ppm perchloroethylene vapor. Consultation with a seasoned veterinary industrial hygienist appears warranted.

254. Three g of liquid mercury were spilled on a carpet. Vacuuming removed the mercury but mercury vapors continued to be emitted from the sweeper's bag. If the evaporation rate during vacuum sweeper operation was 300 micrograms of Hg per min, and the exhaust rate was 50 cfm, what was the concentration of mercury vapor in the air exhausted from the sweeper?

$$\frac{300 \text{ mcg}}{\text{minute}} \times \frac{\text{milligrams}}{1000 \text{ micrograms}} = \frac{0.3 \text{ mg Hg}}{\text{minute}} \qquad \frac{50 \text{ ft}^3}{\text{minute}} \times \frac{M^3}{35.315 \text{ ft}^3} = \frac{1.416 \, M^3}{\text{minute}}$$

$$\frac{0.3 \text{ mg Hg/minute}}{1.416 \, M^3/\text{minute}} = 0.212 \text{ mg Hg/}M^3$$

Answer: 0.212 milligrams of mercury vapor per cubic meter.

255. A paint sprayer using an air line respirator painted the inside of a 14-ft diameter by 30-ft long chemical storage tank with an epoxy resin paint. The paint contained 28% methyl ethyl ketone (by weight) as the solvent. The tank had no ventilation. If one gallon of the coating covers 325 ft², what MEK vapor concentration can be expected in the tank after the painter leaves? The resin coating weighs 12.3 pounds per gallon. MEK's molecular weight = 72.1.

lateral tank surface area = $2 \times \pi \times r \times h = (2) \, \pi \, (7 \text{ ft}) \, (30 \text{ ft}) = 1320 \text{ ft}^2$

end surface areas = $2 \times \pi \times r^2 = (2) \, \pi \, (7 \text{ ft})^2 = 308 \text{ ft}^2$

total tank surface area = $1320 \text{ ft}^2 + 308 \text{ ft}^2 = 1628 \text{ ft}^2$

$$1628 \text{ ft}^2 \times \frac{\text{one gallon}}{325 \text{ ft}^2} = 5.01 \text{ gallons}$$

This was verified by finding an empty five-gallon pail of the resin coating outside of the tank after the painter finished.

$(12.3 \text{ lb/gallon}) \times 5.01 \text{ gallons} \times 0.28 = 17.25 \text{ lb MEK}$

tank volume = $\pi \times r^2 \times h = \pi \, (7 \text{ ft})^2 \, (30 \text{ ft}) = 4618 \text{ ft}^3$

$$4618 \text{ ft}^3 \times 28.32 \text{ L/ft}^3 = 130,782 \text{ L} = 130.78 \text{ M}^3$$

$$\frac{\dfrac{7,832,000 \text{ mg}}{130.78 \text{ M}^3} \times 24.45}{72.1} = 20,308 \text{ ppm MEK vapor}$$

Answer: 20,308 ppm of methyl ethyl ketone vapor (2.03%) is the average concentration in the tank. The LEL = 1.7%. The UEL = 11.4%. Ignition sources? Explosion proof lighting in the dark tank? Blow air into the tank to dilute the vapors to < PEL. Discharge explosive vapors to a safe area. Prohibit entry. Check vapors with a calibrated CGI. Consider diluting MEK vapors with nitrogen at first to get well below the LEL, then follow with air dilution to < PEL.

256. The air flow through a low flow air sampling pump and small charcoal tube was calibrated with a 100 mL soap film bubble tube. The calibration conditions were at NTP. What was the average flow rate in L/min if two bubble traverses required 73.7 and 74.1 sec?

$$\frac{74.1 \text{ seconds} + 73.7 \text{ seconds}}{2} = \frac{73.9 \text{ seconds}}{100 \text{ mL}}$$

$$\frac{60 \text{ seconds}/100 \text{ mL}}{\times \text{ seconds}/100 \text{ mL}} = \frac{0.1 \text{ L}/\text{minute}}{\times \text{ L}/\text{minute}} = \frac{60 \text{ seconds}}{73.9 \text{ seconds}} = 0.812$$

$$\times \frac{0.1 \text{ L}}{\text{minute}} = \frac{0.0812 \text{ L}}{\text{minute}}$$

Answer: 0.081 liter of air sampled per min.

257. An impinger containing 8.9 mL of collection solution was analyzed and found to contain 0.19 mcg TDI/mL. The sample of air had been collected at 1.04 L/min for 16.5 min. If the collection efficiency of the impinger was 85%, what was the concentration of TDI in the breathing zone of the polyurethane foam maker? The molecular weight of TDI = 174.2 g/g-mol.

$$(1.04 \text{ L/min}) \times 16.5 \text{ min} = 17.16 \text{ L}$$

$$8.9 \text{ mL} \times \frac{0.19 \text{ mcg}}{\text{mL}} \times \frac{1}{0.85} = 1.99 \text{ mcg TDI}$$

$$\frac{\dfrac{1.99 \text{ mcg}}{17.16 \text{ L}} \times 24.45}{174.2} = 0.016 \text{ ppm TDI}$$

Answer: 0.016 ppm of toluene 2,4- and 2,6-diisocyanate isomers.

258. Twelve and one-half pounds of methylene chloride evaporate inside of a 50-ft diameter by 95-ft long storage tank. A maintenance worker inspected the tank wearing a HEPA respirator and remained inside the tank for two hrs and 40 min. What was his eight-hr TWAE to this vapor? The molecular weight of $CH_2Cl_2 = 84.9$.

tank volume $= \pi \times (25 \text{ ft})^2 \times 95 \text{ ft} = 186{,}532.5 \text{ ft}^3$

$$186{,}532.5 \text{ ft}^3 \times \frac{28.32 \text{ L}}{\text{ft}^3} \times \frac{M^3}{1000 \text{ L}} = 5282.6 \text{ M}^3$$

$$12.5 \text{ lb} \times \frac{453.59 \text{ g}}{\text{lb}} \times \frac{1000 \text{ mg}}{\text{g}} = 5{,}669{,}875 \text{ mg}$$

$$\frac{\dfrac{5{,}669{,}875 \text{ mg}}{5282.6 \text{ M}^3} \times 24.45 \text{ L/g-mole}}{84.9} = 309 \text{ ppm}$$

160 min $\times$ 309 ppm = 49,440 ppm-min

320 min $\times$ 0 ppm = 0 ppm-min

49,440 ppm-min/480 min = 103 ppm

Answer: HEPA filter respirator is worthless for protection against solvent vapor. His TWAE to CH_2Cl_2 vapor was 103 ppm. ACGIH classifies CH_2Cl_2 as an A2 carcinogen with an eight-hr TWAE TLV of 50 ppm.

In the preceding problem, assume that the worker was found dead in the tank — the unfortunate victim of solvent vapor inhalation. What could explain his death? 309 ppm?

Answer: A 160-min exposure to 309 ppm CH_2Cl_2 would be an unlikely cause of death unless there was ventricular fibrillation by sensitization of the myocardium from endogenously released adrenaline. The solvent's 100% vapor density is much greater than air, and a pocket or stratified vapor layer in the air would have a much greater concentration. The average concentration (309 ppm) can be very misleading when investigating confined spaces where gas and vapor pockets of very high concentrations can occur.

259. A welder was exposed to 260 mcg Pb/M³ for 75 min while burning bolts, 38 mcg Pb/M³ for three hrs and 19 min when welding painted steel, and 560 mcg Pb/M³ for 27 min when using an abrasive blaster to remove lead-containing paint from steel prior to welding. The balance of his 8-hr exposure was 1.5 mcg Pb/M³. What was his TWAE?

job	C (mg Pb/M³)	×	T (minutes)	=	CT (mcg/M³-min)
burning	260	×	75	=	19,500
welding	38	×	199	=	7,562
blasting	560	×	27	=	15,120
other	1.5	×	179	=	269
			480		42,451

$$\frac{42,451 \text{ mcg Pb/M}^3 - \text{minutes}}{480 \text{ minutes}} = \frac{88.4 \text{ mcg Pb}}{\text{M}^3}$$

Answer: Eight-hr time-weighted average exposure to lead dust and fume = 88 mcg Pb/M³. This exceeds OSHA's PEL by about 77%. Implement the OSHA lead standard (29 *CFR* 1910.1025).

260. A 10-pound chunk of calcium carbide fell into an open 55-gallon drum of water in a room with dimensions of 10 × 12 × 12 ft. The room had no ventilation, but there was a candle burning in a corner. There was quantitative conversion of the calcium carbide to acetylene gas. Would there have been an explosion? Disregard any solubility of the acetylene in water. The respective molecular weights of calcium carbide and acetylene are 64.1 and 26.04.

$CaC_2 + 2 H_2O \rightarrow Ca(OH)2 + C_2H2\uparrow$

10 lb × 453.59 g/lb = 4535.9 g of calcium carbide

$$\frac{4535.9 \text{ grams CaC}_2}{64.1 \text{ grams/mole}} = 70.76 \text{ moles of CaC}_2$$

Therefore, 70.76 mols of acetylene gas were generated and released to the air.

70.76 mols C_2H_2 × 26.04 g/mol = 1842.67 g C_2H_2

10 × 12 × 12 ft = 1440 ft³ × (28.32 L/ft³) = 40,781 L = 40.78 M³

$$\frac{\dfrac{1,842,670 \text{ mg}}{40.78 \text{ M}^3} \times 24.45 \text{ L/g-mole}}{26.04 \text{ g/g-mole}} = 42,427 \text{ ppm acetylene gas}$$

Answer: The average concentration of acetylene gas in the room is 4.24%. The LEL–UEL range for acetylene gas in air is one of the widest of all explosive hydrocarbon gases: 2.5% to 80%. Although the concentration is toward the lower end of the explosive range for acetylene, this room will be blown to kingdom come.

261. A railroad tank car containing nine tons of liquid chlorine derails, tips, ruptures, and spills its contents into a drainage ditch which is parallel to the right-of-way. A very light, hot summer breeze evaporates some liquid into a gas cloud which is about 1.3 miles long, 0.67 mile wide, and 200 ft high. What

is the average Cl_2 gas concentration in this cloud? Assume 20% of the liquid chlorine in the ditch volatilizes before the remaining liquid chlorine is contained.

1.3 miles $\times$ (5280 ft/mile) = 6864 ft   0.67 mile $\times$ (5280 ft/mile) = 3538 ft

200 $\times$ 6864 $\times$ 3538 ft = 4.857 $\times$ 10^9 ft³ $\times$ (28.32 L/ft³) = 1.376 $\times$ 10^{11} L = 1.376 $\times$ 10^8 M³

$$9 \text{ tons} \times \frac{2000 \text{ lb}}{\text{ton}} \times \frac{453.59 \text{ grams}}{\text{lb}} \times \frac{1000 \text{ mg}}{\text{gram}} \times 0.20 = 1.633 \times 10^9 \text{ mg } Cl_2$$

$$\frac{\dfrac{1.633 \times 10^9 \text{ mg } Cl_2}{1.376 \times 10^8 \text{ M}^3} \times \dfrac{24.45 \text{ L}}{\text{g-mole}}}{70.9} = 4.1 \text{ ppm } Cl_2$$

Answer: The average concentration of chlorine gas in this cloud is 4.1 ppm. However, while the calculation is helpful for air pollution and community risk assessment purposes, a much greater concentration exists close to the spill which is subject to the prevailing meteorological conditions. This would help determine the gas diffusion parameters in x, y, and z planes and the expected concentrations at selected points away from the spill.

262. Air was to be sampled through a midget impinger containing 15 mL distilled water for a specified time at 1.09 L/min. The barometric pressure is 710 mm Hg. The ambient temperature is 14°F. The laboratory analytical detection limit is 320,000 particles per mL. How long should one sample to determine 20% of a 30 mppcf dust concentration?

Answer: Sample until the water freezes in the impinger (which will happen in a minute or so). Add NaCl, MeOH, or IPA to lower the freezing point of the liquid. If this is done, the following calculations apply:

$$\frac{320,000 \text{ particles}}{\text{mL}} \times 15 \text{ mL} = 4,8000,000 \text{ particles}$$

$$20\% \text{ of } 30 \text{ mppcf} = \frac{6 \times 10^6 \text{ particles}}{\text{ft}^3} \times \frac{\text{ft}^3}{28.32 \text{ L}} = \frac{211,864 \text{ particles}}{\text{liter}}$$

$$\frac{4,800,000 \text{ particles}}{211,864 \text{ particles/L}} = 22.66 \text{ L} \qquad \frac{12.6 \text{ L}}{1.09 \text{ L/minute}} = 20.7 \text{ minutes}$$

263. The formula Q = 0.75 V (10 X^2 + A) is used for calculations involving which type of exhaust hood?

    a. Slot hood without a flange and an aspect ratio of 0.2 or less.
    b. Slot hood with flange and aspect ratio of 0.2 or less.

    c. Plain opening hood without flange with aspect ratio of 0.2 or greater or round without flange.

    d. Plain opening hood with flange and aspect ratio of 0.2 or greater or round hood without flange.

    e. Circular canopy hood for hot gases, smoke, fumes, steam.

    f. Plain multiple slot opening hood with two or more slots and an aspect ratio of 0.2 or greater.

    g. All sheet metal hoods, but not for flexible duct hoods.

Answer: d.

264. What is the expected capture velocity eight in. in front of a 2.5-in × 3-ft flanged slot hood with a slot velocity of 2000 fpm?

The basic equation for air flow for a flanged slot with an aspect ratio of 2.5/36 in. = 0.07 is Q = 2.6 LVX, where: Q = air volume rate, L = slot length, V = capture velocity, and × = the distance in front of the hood face.

$$Q = A \times V = \frac{2.5'' \times 36''}{144 \text{ in}^2/\text{ft}^2} \times \frac{2000 \text{ feet}}{\text{minute}} = \frac{1250 \text{ ft}^3}{\text{minute}}$$

$$V = \frac{Q}{2.6 \times L \times X} = \frac{1250 \text{ cfm}}{2.6 \times 3 \text{ ft} \times 0.67 \text{ ft}} = \frac{239 \text{ f}}{\text{minut}}$$

Answer: 239 ft/min.

265. About five g of mothball crystals (*p*-dichlorobenzene) remain in an empty dresser drawer after sweaters had been removed. The drawer is 19 × 11 × 28 in. The molecular weight of DCB is 147. DCB vapor pressure is 0.4 mm Hg at 75°F. What is the vapor concentration in the closed drawer expressed in mg/M³? How many milligrams of DCB vapor are in the air of the drawer? The OSHA PEL for DCB is 75 ppm. Assuming toxicokinetics of DCB in humans are similar to those in a moth, is the moth excessively exposed?

$$\frac{0.4 \text{ mm Hg}}{760 \text{ mm Hg}} \times 10^6 = 526 \text{ ppm of DCB vapor at saturation}$$

$$\frac{\text{mg}}{\text{M}^3} = \frac{\text{ppm} \times \text{molecular weight}}{24.45 \text{ L/g-mole}} = \frac{526 \text{ ppm} \times 147}{24.45} = \frac{3162 \text{ mg DCB}}{\text{M}^3}$$

$$\frac{\frac{19'' \times 11'' \times 28''}{1728 \text{ in}^3}}{\text{ft}^3} \times \frac{28.32 \text{ L}}{\text{ft}^3} \times \frac{\text{M}^3}{1000 \text{ L}} = 0.0959 \text{ M}^3$$

$$\frac{3162 \text{ mg}}{\text{M}^3} \times 0.0959 \text{ M}^3 = 303 \text{ milligrams of DCB}$$

Answers: 3162 mg DCB/M³. 303 mg DCB vapor. Yes.

266. The velocity pressure in a duct exhausting standard air is 0.84-in wg. What is the duct air velocity?

$$V = 4005 \times \sqrt{VP} = 4005 \times \sqrt{0.84} = 3671 \text{ fpm}$$

Answer: 3671 ft/min.

267. An exhaust duct static pressure is 1.34 in. wg. The total system pressure is –0.63 in. wg. What is the standard air velocity pressure and the duct velocity?

Since this is an exhaust duct, the static pressure is negative, or –1.34 in. wg.

$$SP + VP = TP \quad VP = TP - SP$$

$$VP = -0.63 \text{ in.} - (-1.34 \text{ in.}) = 0.71 \text{ in.} \; \textit{Note:} \text{ VP is always positive.}$$

$$V = 4005 \times \sqrt{VP} = 4005 \times \sqrt{0.71} = 3375 \text{ feet per minute}$$

Answer: velocity pressure = 0.71-in wg. Duct velocity = 3375 fpm.

268. A 14-in diameter duct has an average duct velocity of 3160 fpm. What is the duct's volumetric flow rate?

$$Q = A \times V = \frac{\pi \times (7'' \times 7'')}{144 \text{ in}^2/\text{ft}^2} \times \frac{3160 \text{ ft}}{\text{minute}} = \frac{3378 \text{ ft}^3}{\text{minute}}$$

Answer: 3378 cubic ft of air/min.

269. Standard dry air at 70°F and 29.92-in barometric pressure has a mass of:

   a. 0.13 lb/ft³
   b. 0.013 lb/ft³
   c. 22.4 lb/1000 ft³
   d. 0.003 grains/ft³
   e. 0.130 lb/1000 ft³
   f. 0.075 lb/ft³

Answer: f.

270. The LEL for most combustible gases and vapors is normally constant up to about 250°F. What approximation factor is usually applied above these temperatures to account for the change in the LEL ?

   a. 0.3
   b. 0.5
   c. 0.7

    d. 0.9

    e. 1.3

    f. 1.5

Answer: c.

271. A 47 × 166 × 20-ft building is supplied with 7300 cfm. How many air changes occur per hr? How many minutes are required per air change? How many cubic feet are supplied per square foot of floor area? The ceiling height is 20 ft. Forty-seven people work in this single-story building. What is the outdoor air ventilation rate per person if 90% of the air is recirculated?

7300 cfm × 60 min/hr = 438,000 cfh

$$\frac{438,000 \text{ cfh}}{47 \times 166 \times 20 \text{ ft}} = 2.8 \text{ air changes/hour}$$

$$\frac{60 \text{ minutes/hour}}{2.8 \text{ air changes/hour}} = 21.4 \text{ minutes/air change}$$

$$\frac{7300 \text{ cfm}}{47 \text{ ft} \times 166 \text{ ft}} = 0.94 \text{ cfm/1}$$

$$\frac{0.1 \times 7300 \text{ cfm}}{47 \text{ occupants}} = \frac{15.5 \text{ ft}^3/\text{minute}}{\text{occupant}}$$

Answers: 2.8 air changes per hr. 21.4 min per air change. 0.94 cfm per ft² of floor area. 15.5 cubic ft of outside air/min per worker/occupant.

272. Industrial hygienists rarely prescribe canopy hoods because of their often poor performance, and because these hoods are vulnerable to cross drafts. One engineering equation used to design such hoods, however, is:

    a. $Q = 1.4 \text{ PDV}$

    b. $Q = (\text{LWH}/2) \times V$

    c. $Q = C_e \times SP_h \times V$

    d. $Q = V (5 \ x^2 + A)$

    e. $Q = \pi \text{ CLW}$

Answer: a. Where P = tank perimeter, in feet. D = vertical distance between tank edge and bottom edge of canopy, in feet. V = design capture velocity (fpm), which can range from as low as 50 fpm in undisturbed locations to 500 fpm or higher when there are moderate cross drafts and thermal convection air currents.

273. If 1000 cfm are needed to capture air contaminants six in. from a hood's face, what exhaust air volume is needed to capture contaminants 12 in. from the hood's face?

    a. 500 cfm
    b. 1000 cfm
    c. 2000 cfm
    d. 4000 cfm
    e. 9000 cfm
    f. 16,000 cfm

    Answer: d. The required air volume varies as the square of the distance from the air contaminant source. This points out the importance of placing the exhaust hood face as close as possible to the source of air pollutants. Exhaust volume is conserved, and more effective capture is achieved.

274. A side draft exhaust hood with face dimensions of 32 × 50 in. has an average face velocity of 110 fpm. What volume of air is exhausted by this hood? What will be the average face velocity if the hood opening is reduced to 26 × 40 in?

$$\frac{32"\times 50"}{144 \text{ in}^2/\text{ft}^2}\times 110 \text{ fpm} = \frac{1222 \text{ ft}^3}{\text{minute}}$$

$$\text{new area} = \frac{26"\times 40"}{144 \text{ in}^2/\text{ft}^2} = 7.2 \text{ ft}^2$$

$$V = Q \times A = \frac{1222 \text{ cfm}}{7.2 \text{ ft}^2} = 170 \text{ fpm}$$

    Answer: 1222 cfm. The hood capture velocity increased to 170 fpm by slightly reducing the opening and improving the enclosure.

275. The static pressure at the fan inlet on an exhaust system is –2.75-in H_2O. Static pressure at the fan outlet is 0.25 in. The diameter of the exhaust duct at the fan inlet is an odd size: 23 in. The system is exhausting 4000 cfm. What is the fan static pressure?

    23-in diameter duct = 2.885 ft²

$$\text{fan inlet velocity, V} = \frac{4000 \text{ cfm}}{2.885 \text{ ft}^2} = \frac{1836 \text{ ft}}{\text{minute}}$$

$$VP = \frac{V^2}{(4005)^2} = \frac{(1386)^2}{(4005)^2} = 0.12"$$

    fan SP = SP_{in} + SP_{out} – VP_{in} = 2.75 in. + 0.25 in. – 0.12 in. = 2.88 in

Answer: 2.88 in. of water. This figure is used to help select the appropriate fan from commercial fan tables and catalogues.

276. An exhaust system operates at 19,400 cfm. A hood is added to the system which requires a total system capacity of 23,700 cfm. By how much should fan speed be increased to handle the extra exhaust volume? Let Q_1 = original exhaust volume, and Q_2 = new exhaust volume.

Since CFM varies directly with fan rpm: $\dfrac{Q_2}{Q_1} = \dfrac{rpm_2}{rpm_1} = \dfrac{23,700 \text{ cfm}}{19,400 \text{ cfm}} = 1.22.$

Answer: Increase fan speed by 22%, i.e., multiply the fan rpm $\times$ 1.22. *Caution:* Ensure that the maximum fan rpm and safety-rated tip speed are not exceeded.

277. In the preceding problem, what is the required increase in the fan horsepower to handle the increased air volume?

Fan horsepower varies as the cube of the fan speed and cube of the fan volume. Therefore, the required ratio of increase in the fan speed (rpm) is:

$$\left[\frac{23,700 \text{ cfm}}{19,400 \text{ cfm}}\right]^3 = 1.82$$

Answer: The required ratio of the increase in the fan horsepower is 1.82 — an 82% increase in the energy requirement for only a 22% increase in fan volume and speed. *Caution:* Be certain the fan horsepower rating is ample. Check the motor plate for specifications and maximum operating capacity. Consult with a professional electrician and a millwright.

278. A ventilation system was designed to exhaust 17,500 cfm at a static pressure of 3.46 in. at the fan inlet. A static pressure reading at the fan inlet is 2.66 in. two years after the system was installed. What is the present exhaust volume? What's up?

$$Q_{now} = Q_{then} \times \sqrt{\frac{SP_{now}}{SP_{then}}} = 17,500 \text{ cfm} \times \sqrt{\frac{2.66''}{3.46''}} = 17,500 \text{ cfm} \times 0.877 =$$

15,348 cfm   17,500 cfm − 15,348 cfm = 2152 cfm

Answer: The system volume decreased 2152 cfm, or about 12%. Diagnosis could include dirty filters and ducts, slipping fan belt, sticking dampers, corrosion and holes in the system (air shunting), and plugged and dirty heat exchange coils. *Note:* The exhaust volume varies as the square of the static pressure at the fan inlet, whereas the exhaust volume varies directly as the fan revolutions and as the cube of the horsepower.

279. What is the discharge air volume from a 50 ft high plant with a 30 ft² outlet in the roof and a 35 ft² inlet at the ground level if the outdoor temperature is 20°F and the discharge air temperature is 75°F (Δ t = 55°F)?

ASHRAE provides a convenient formula to estimate the "chimney" or "stack effect" ventilation due to temperature differentials existing between the floor of a building and its ceiling:

$$Q(\text{cfm}) = 9.4 \times A \times h \times \sqrt{t_{in} - t_{out}}, \text{ where:}$$

A = total inlet or discharge area (ft²) at ground or ceiling
      (use the smaller of the two)
h = distance between inlet and outlet, ft
t_{in} = air inlet temperature, °F
t_{out} = air outlet temperature, °F

$$Q = 9.4 \times 30 \text{ ft}^2 \times 50' \times \sqrt{75°F - 20°F} = 14,788 \text{ cfm}$$

Answer: About 14,800 cfm are naturally exhausted due to the thermal gradient. This demonstrates that large volumes of natural ventilation result from differences in temperature forces alone. For example, it has been estimated that, in some steel mills, 600 tons of air are naturally exhausted for every ton of steel produced.

280. A 10-in diameter circular air supply duct is blowing air at 4200 ft/min as measured at the discharge point. What general quantitative statement applies regarding air velocity ("throw") at some distance away from the duct outlet?

Answer: The velocity 30 duct diameters away is about 10% of the discharge velocity. That is, in this case, 30 duct diameters = 30 × 10 in. = 300 in. = 25 ft. The air velocity at this distance reduces to nearly 420 fpm if there are no significant cross drafts.

281. A low circular canopy hood is positioned over a circular tank containing water at 180°F. The distance between the tank and hood is less than three ft. If the room air temperature is 70°F and the tank diameter is four ft, what is the total air flow rate required for this hood to capture the water vapor?

Q = 4.7 × (D)$^{2.33}$ × (T)$^{0.42}$, where:
Q = total hood air flow, cfm
D = hood diameter, feet (add one foot to source diameter)
T = difference between the temperature of the hot source of air contaminants
      and the ambient temperature, °F
Q = 4.7 × (4' +1')$^{2.33}$ × (180°F–70°F)$^{0.42}$ = 4.7 × (5)$^{2.33}$ × (110)$^{0.42}$ = 4.7 ×
      36.2 × 7.2 = 1225 cfm

Answer: 1225 cubic ft of air/min. Fan selection should be based on the thermally expanded air volume and the air temperature existing at the inlet of the fan.

282. What is the approximate capture velocity 12 in. away from a one-foot diameter duct exhausting air at a duct velocity of 3600 fpm?

    Answer: A rule of thumb is that the capture velocity one duct diameter away from an exhaust opening is about 10% of the duct velocity or, in this case, 360 fpm.

283. To ensure uniform capture and air distribution which is applicable to most exhaust hoods, a maximum plenum velocity should be no more than _____ % of the slot velocity.

    a. 10%
    b. 30%
    c. 50%
    d. 75%
    e. 110%
    f. 150%

    Answer: c. For example, the maximum plenum velocity for a 2000 fpm slot hood should not exceed 1000 fpm. This is a good choice for uniform air flow across the slot and a moderate pressure drop. If, during design, the calculated plenum velocity is too high (i.e., > 50% of the design slot velocity), the size of the plenum must be increased.

284. A standard bench grinder hood has a coefficient of entry of 0.78. A piece of string wrapped around the take-off exhaust duct to measure the circumference was 16 in. The throat static pressure of the hood is –2.75 in. What volume of air is exhausted by this hood?

    $$Q\,(\text{cfm}) = 4005 \times A \times C_e \times \sqrt{SP}, \text{ where:}$$

    A   = cross section area of duct at throat, ft^2
    C_e = coefficient of hood entry
    SP = static pressure at hood throat, inches of water

    $$\text{duct circumference} = 2 \times \pi \times r = 16" \qquad r = \frac{16"}{2 \times \pi} = 2.55 \text{ inches}$$

    Therefore, the internal duct diameter = 5 in. (a standard size — the true duct radius = 2.5 in).

$$A = \frac{\pi \times (2.5")^2}{\dfrac{144 \text{ in}^2}{\text{ft}^2}} = 0.136 \text{ ft}^2$$

Q = 4005 × 0.136 ft² × 0.78 × (2.75)⁰·⁵ = 705 cfm

Answer: This grinder exhausts 705 cubic ft of air/min.

285. In general, the addition of a flange to an exhaust system hood opening improves air contaminant capture by about _____ percent.

    a. 15%
    b. 25%
    c. 40%
    d. 60%
    e. 75%

Answer: b. Alternatively, at equal capture velocities, the flange decreases the required air flow rate by about 25%. Typically, the width of the flange should equal or exceed 25% of the square root of the hood's face area. In other words, a hood 5 × 6 ft = 30 ft². 0.25 [(30 ft²)⁰·⁵] ≅ 1.4 ft.

286. A duct velocity of 4005 ft/min is equivalent to what velocity pressure?

The basic equation is: $V = 4005 \times \sqrt{VP}$.

$$\frac{4005}{4005} = \sqrt{VP} \quad VP = 1.0"$$

Answer: The velocity pressure equals one in of water assuming dry air at standard temperature.

287. A ventilation system supplies air through five branch ducts. The system delivers 15,000 cfm and another duct will be added to increase the total supply volume to 17,000 cfm. The static pressure at the fan discharge is 0.59 in. The static pressure in the main duct where the branch is to be added is 0.38 in. What are the new static pressures at the fan discharge and at the fan inlet plenum?

Calculate the new SP drop in the main duct from the fan to the new connection:

$$\sqrt{\frac{17,000 \text{ cfm}}{15,000 \text{ cfm}}} \times (0.59" - 0.38") = 1.065 \times 0.21 = 0.22" \text{ H}_2\text{O}$$

new SP at fan discharge = 0.38 in. H₂O + 0.22 in. H₂O = H₂O

$$\text{new static pressure at fan inlet plenum} = \frac{17,000 \text{ cfm}}{15,000 \text{ cfm}} \times 0.30" = 0.34" \text{ H}_2\text{O}$$

Answers: 0.22 in. = the new static pressure drop at new duct. 0.50 in. = new static pressure at the fan discharge. 0.34 in. = the new static pressure at the fan inlet. The required increase in fan rpm and horsepower can now be obtained from fan performance catalogues.

288. A five horsepower fan motor is exhausting 3250 cfm. Amperes measured on the circuit are 9.8. The fan motor's rated amps are 13.2 for five horsepower at 220 volts. What will be the new air volume if the fan speed is increased to fully load the motor at five horsepower?

$$\text{actual horsepower} = \frac{\text{measured amperage}}{\text{rated amperage}} \times \text{rated horsepower} = \frac{9.8 \text{ amps}}{13.2 \text{ amps}} \times$$

5 HP = 3.7 horsepower

According to the fan laws, horsepower varies as the cube of the air volume (cfm) or the fan speed (rpm):

$$\frac{HP_1}{HP_2} = \left[\frac{Q_1}{Q_2}\right]^3 \qquad Q_2 = Q_1 \times \left[\frac{HP_2}{HP_1}\right]^{0.333} = 3250 \text{ cfm} \times \sqrt[3]{\frac{5 \text{ HP}}{3.7 \text{ HP}}} = 3592 \text{ cfm}$$

Answer: The new volume at full five horsepower = 3592 cubic ft/min.

289. What is the density of dry air at 120°F? The density of air at 70°F = 0.75 lb/ft³.

Air density varies with the absolute temperature. That is, the density decreases as the air temperature increases:

$$\frac{0.075 \text{ lb}}{\text{ft}^3} \times \frac{460° + 70°\text{F}}{460° + 120°\text{F}} = \frac{0.075 \text{ lb}}{\text{ft}^3} \times \frac{530°\text{A}}{580°\text{A}} = \frac{0.0686 \text{ lb}}{\text{ft}^3}$$

Answer: The density of dry air at 120°F = 0.0686 lb/ft³.

290. In problem 289, what is the density of the air at 2000 ft elevation? The standard barometric pressure at 2000 ft is 27.80 in. of Hg.

The density of air (or any gas or vapor) is reduced by elevation:

$$\frac{0.075 \text{ lb}}{\text{ft}^3} \times \frac{530°\text{A}}{580°\text{A}} \times \frac{27.80" \text{ Hg}}{29.92" \text{ Hg}} = \frac{0.0637 \text{ lb}}{\text{ft}^3}$$

Answer: 0.0637 lb/ft³ of dry air. Therefore, it can be seen that as air is heated and reduced in pressure its density is reduced by both forces. That is, air expands into a larger volume but retains its original mass.

291. What happens to air density when an inlet duct is used? For example, calculate the density correction for an inlet duct with a 20-in inlet suction at an atmospheric pressure of 407 in. Apply this to air at 120°F and 2000 ft elevation.

$$\frac{407" - 20"}{407"} = 0.951$$

$$density = \frac{0.075\ lb}{ft^3} \times \frac{530°A}{580°A} \times \frac{27.80"\ Hg}{29.92"\ Hg} \times \frac{0.951 \times 407"}{407"} = \frac{0.0605\ lb}{ft^3}$$

Answer: 0.0605 lb/ft³.

292. What happens to the density of dry air as water vapor is added to it?

Answer: The density of air is reduced as water vapor is added to it (as in wet scrubber systems) since the water vapor molecules weigh less than air molecules ("apparent" molecular weight of air = 29; molecular weight of water = 18). As an example, consider water-saturated air at 120°F, 2000 ft elevation, and at 20 in. suction (See problems 289, 290, and 291):

$$\frac{0.0605\ lb}{ft^3} \times \frac{27.80"\ Hg}{29.92"\ Hg} \times \frac{387"}{407"} = \frac{0.0535\ lb}{ft^3}$$

293. A fan must be selected for a 20 in. suction pressure at an air density of 0.0605 lb/ft³. How would you do this?

Since the pressure ratings of fans are based upon a standard gas density of 0.075 lb/ft³, the selection pressure must be adjusted to a density of 0.075 lb/ft³.

$$fan\ selection\ static\ pressure = 20" \times \frac{0.075\ lb/ft^3}{0.0605\ lb/ft^3} = 24.8"$$

Answer: 24.8 in. of water static pressure.

294. The minimum basic design ventilation rates, respectively, for propane fuel and gasoline fuel lift trucks are:

a. 300 cfm and 600 cfm
b. 600 cfm and 300 cfm
c. 1000 cfm and 2000 cfm
d. 6000 cfm and 10,000 cfm
e. 5000 cfm and 8000 cfm

Answer: e. This assumes a regular engine maintenance program, less than 1% and 2% CO in the exhaust gas, truck operating times less than 50%, good distribution of dilution ventilation air flow, and at least 150,000 ft³ plant volume per lift truck. See ACGIH's *Industrial Ventilation* for other conditions of maintenance, truck operation, and ventilation parameters.

295. A fan exhausts 1000 cfm of air at 600°F at a static pressure of 10 in. of H_2O. What is the required fan horsepower? What will be the horsepower savings (that is, electricity costs) if the air is cooled to 100°F?

fan horsepower = 0.000158 × 1000 cfm × 10 in. H_2O = 1.58

Use Charles' law to calculate the new air volume at 100°F:

$$\frac{Q\,(at\ 100°F)}{Q\,(at\ 600°F)} = \frac{100°F + 460°}{600°F + 460°}$$

$$Q\,(at\ 100°F) = \frac{1000\ ft^3}{minute} \times \frac{560°A}{1060°A} = 528\ cfm,\ where$$

Q (at 100°F) denotes the volume flow rate at 100°F, and (100°F + 460°) is the absolute temperature. Assuming the increase in density at the lower temperature results in a negligible reduction in pressure loss, then at 100°F:

fan horsepower = 0.000158 × 528 cfm × 10 = 0.83 HP

Answer: 0.83 horsepower which = (0.83/1.58) × 100 = 52.5% of the horsepower required at 600°F.

296. A paint spray booth is seven ft high × 10 ft wide. Spraying is often done as far as five ft in front of the booth. A nearly draft-free area requires 100 ft/min capture velocity at the point of spraying. Determine the required exhaust rate.

$$Q = V\left[\frac{10x^2 + 2A}{2}\right]$$

This is a modification of the classic ventilation equation which applies to a free-standing, unobstructed hood. The above equation is applied to rectangular hoods bounded on one side by a plane surface. The hood is considered to be twice its actual size with an additional portion being the mirror image of the actual hood and the bounding plane being the bisector.

$$Q = 100\ fpm \times \frac{\left[(10)(5\ ft)^2\right] + (2)(7\ ft)(10\ ft)}{2} = \frac{19,500\ ft^3}{minute}$$

$$hood\ face\ velocity = \frac{Q}{A} = \frac{\dfrac{19,500\ ft^3}{minute}}{7 \times 10\ feet} = \frac{280\ feet}{minute}$$

Answer: exhaust rate = 19,500 cfm. Face velocity = 280 fpm.

297. A 28-in diameter fan operating at 1080 rpm supplies 4700 cfm at 4.75 in. of static pressure. What size fan of the same type and series would supply 10,900 cfm at the same static pressure?

$$\text{fan diameter} = \left[\frac{10,900 \text{ cfm}}{4700 \text{ cfm}}\right]^{0.333} \times 28" = 37 \text{ inches}$$

Answer: 37 in. diameter, an odd dimension. A 36-, 38- or 40-in fan most likely would have to be selected from those available from most manufacturers.

298. A fan exhausts 600°F from a drier at 12,000 cfm. The density of this air is 0.0375 lb/ft³ (because of its high water vapor content). The static pressure at the dryer's discharge is four in of water. The fan speed is 630 rpm. The fan uses 13 horsepower. What is the required horsepower if 70°F air is exhausted?

$$HP = 13 \text{ HP} \times \frac{0.075 \text{ lb/ft}^3}{0.0375 \text{ lb/ft}^3} = 26 \text{ HP}$$

Answer: 26 horsepower is required. If only a 15 horsepower fan, for example, was in this system requiring 26 HP, it would be necessary to use a damper when starting "cold" to prevent electrically overloading the fan motor. Always ensure the installation of a fan with sufficient horsepower to handle the system requirements. The cost of installing a slightly larger fan is often justified at initial installation since future system modifications can be anticipated (e.g., adding another hood to the exhaust system or a collector with a higher static pressure drop).

299. Assume that there are 63,000,000 automobiles in the United States and that each uses an average of three gallons of gasoline every day. Further assume there are no gasoline vapor recovery systems when the car's fuel tank is filled and that each gallon of gasoline pumped to the fuel tank displaces a gallon of saturated gasoline vapor. If the average molecular weight of gasoline is taken as approximately 72, and the average vapor pressure of gasoline is 130 mm Hg, how many tons of gasoline vapor are evaporated and enter the atmosphere by filling automobile fuel tanks every year in the U.S.? Assume the national average barometric pressure for high density population areas is 740 mm Hg and that the average temperature for these areas is 50°F.

$$\frac{130 \text{ mm Hg}}{740 \text{ mm Hg}} \times 10^6 = 175,676 \text{ ppm gasoline vapor at saturation}$$

$$\frac{\text{mg}}{\text{M}^3} = \frac{\text{ppm} \times \text{molecular weight}}{23.85 \text{ L/g-mole at 50°F}} = \frac{175,676 \times 72}{23.85} = \frac{530,343 \text{ mg}}{\text{M}^3} = \frac{530.34 \text{ g}}{\text{M}^3}$$

$$\frac{530.34\ g}{M^3} \times \frac{lb}{454\ g} \times \frac{M^3}{35.315\ ft^3} \times \frac{0.1337\ ft^3}{gallon} = \frac{0.0044225\ lb}{gallon}$$

$$\frac{0.0044225\ lb}{gallon} \times \frac{3\ gallons/day}{automobile} \times \frac{365.25\ days}{year} \times 63,000,000\ automobiles =$$

$$\frac{3.053 \times 10^8\ lb}{year} \times \frac{ton}{2000\ lb} = \frac{152,648\ tons}{year}$$

Answer: Approximately 150,000 tons of gasoline vapor per year. This crude estimate does not consider evaporative hydrocarbon losses from trucks, airplanes, locomotives, etc. Nor are the hydrocarbon emissions which result from careless overfilling of tanks, refining, filling storage tanks and trucks, etc. considered in this simplified accounting.

300. An analytical chemist found that the average concentration of butyric acid in an adult tennis shoe was 4.7 micrograms after worn in a basketball game. What is the concentration of butyric acid existing in an unventilated gymnasium locker after evaporating from a pair of shoes? The locker is 12 × 12 × 66 in. Molecular weight of butyric acid = 88.11.

$$\frac{4.7\ mcg\ BA}{shoe} \times two\ shoes = 9.4\ micrograms\ of\ butyric\ acid\ vapor$$

$$\frac{12'' \times 12'' \times 66''}{\dfrac{1728\ in^3}{ft^3}} \times \frac{28.32\ L}{ft^3} = closed\ locker\ volume = 155.8\ L$$

$$ppm = \frac{\dfrac{mcg}{L} \times \dfrac{24.45\ L}{g\text{-}mole}}{88.1\ g/g\text{-}mole} = \frac{\dfrac{9.4\ mcg}{155.8\ L} \times 24.45}{88.1} = 0.0167\ ppm\ BA$$

Answer: 16.7 parts of butyric acid vapor per billion parts of stale locker air. The odor threshold for butyric acid (0.0006 ppm = 0.6 ppb) — a stench — is considerably less (found in rancid butter and putrefying animal fatty acids).

301. Silica sand and steel shot are used in an abrasive blasting cabinet to remove lead-containing paint, cadmium plating, plutonium-mercury-beryllium alloy, and osmium tetroxide from arsenic-nickel castings. The interior dimensions of this cabinet are 3.5 × 3.5 × 4.5 ft. The total area of openings into the cabinet is 1.7 ft². What is the required exhaust rate if no less than a 500 fpm face capture velocity is used? What is the exhaust rate if no less than 20 air changes per min are necessary in the cabinet?

$$Q = A \times V = 1.7\ ft^2 \times 500\ fpm = 850\ cfm$$

Q required for 20 air changes/min = 20 × booth volume = 20 × (3.5 × 3.5 × 4.5 ft) = 1103 cfm.

Recalculating face capture velocity using the larger of the two exhaust volumes:

V = Q/A = 1103 cfm/1.7 ft² = 649 fpm

Answers: 850 cfm and 1103 cfm. Use the larger of these. Give very careful attention to the engineering control of dust emissions in the exhaust air. Perhaps the use of ultra-high efficiency bag filters connected in series (and in parallel for system maintenance) should be considered. Maybe the bag house filter housings should be in separate exhausted enclosures with their independent dust collection and absolute containment systems.

302. Two duct branches in an exhaust ventilation system have greatly different static pressures at their union. Balance could be achieved by increasing the flow in the branch with the lower loss. The air volume for the smaller branch is 580 cfm. The static pressure calculated for the branch is −1.75 in. The static pressure for the combined branches at the junction is −3.5 in. What is the corrected exhaust air volume for this branch?

Since pressure losses increase as the square of the volume, increase the air flow through the branch with the lower resistance:

$$Q_{new} = Q_{calc} \times \sqrt{\frac{SP_{junction}}{SP_{branch}}} = 580 \text{ cfm} \times \sqrt{\frac{-3.5''}{-1.75''}} = 580 \text{ cfm} \times \sqrt{2} = \frac{820 \text{ ft}^3}{\text{minute}}$$

Answer: The new ventilation volume rate in the smaller branch is 820 cfm.

303. A sulfur-bearing fuel oil is burned in a combustion process using 20% excess air. Analysis of the oil is (% by weight): 88.3% carbon, 9.5% hydrogen, 1.6% sulfur, 0.10% ash, and 0.05% water. Assuming 4% conversion of sulfur dioxide to sulfur trioxide, what is the required amount of combustion air and the total concentration of sulfur oxides in the stack flue gases? What is the ash concentration in the flue gases at 12% CO_2 when assuming complete combustion? This fuel oil has a theoretical dry air combustion requirement of 176.3 standard cubic feet (scf) per pound. The combustion requirement is 177.6 scf at 40% relative humidity and 60°F per pound of this fuel oil.

Use one pound of this fuel as the basis for all calculations.

Theoretical combustion air requirement:

$C + O_2 \rightarrow CO_2\uparrow$ (0.883 lb) (32/12) = 2.35 lb of O_2
$H_2 + 1/2\ O_2 \rightarrow H_2O\uparrow$ (0.095 lb) (16/2) = 0.076 lb O_2
$S + O_2 \rightarrow SO_2\uparrow$ (0.016) (32/32)   = 0.016 lb O_2

2.35 lb O_2 + 0.76 lb O_2 + 0.016 lb O_2 = 3.13 lb O_2 per pound for CO_2:

$$\frac{2.35\ lb}{3.13\ lb} \times \frac{176.3\ ft^3}{lb\ oil} = 132.4\ scf\ of\ air$$

for H_2:

$$\frac{0.76\ lb}{3.13\ lb} \times \frac{176.3\ ft^3}{lb\ oil} = 42.8\ scf\ of\ air$$

for S:

$$\frac{0.016\ lb}{3.13\ lb} \times \frac{176.3\ ft^3}{lb\ oil} = 0.90\ scf\ of\ air$$

air requirements at 20% excess combustion air:

176.3 scf $\times$ 1.2 = 212 scf dry air/lb of oil
177.7 scf $\times$ 1.2 = 213 scf moist air/lb of oil

combustion products:

(0.883 lb CO_2) (44/12) (379 scf/44 lb/mol) = 27.9 scf CO_2

(0.095 lb H_2O) (18/2) (379 scf/18 lb/mol) = 18.0 scf H_2O from combustion

(0.0005 lb H_2O) (379 scf/18) = 0.011 scf from water in fuel

nitrogen: (212 scf) (0.79) = 167.5 scf

water in air at 40% relative humidity and 60°F:

(0.0072 scf/scf air) (213 scf) = 1.5 scf H_2O

sulfur oxides as SO_2: (0.016) (64/32) (379/64) = 0.19 scf SO_2

oxygen: (176.3 scf) ((1.20 − 1.00) (0.21) = 7.4 scf

total scf of combustion products: 27.9 + 18.0 + 0.011 + 167.5 + 1.5 + 0.19 + 7.4 = 222.3 scf

SO_2 concentration:

(0.016) (379/32) (1/222.3) (10^6) (0.96) = 818 ppm SO_2

SO_3 concentration:

(0.016) (379/32) (1/222.3) (10^6) (0.04) = 34 ppm SO_3

flue ash concentration:

(0.001 lb) (7000 grains/lb) (1/222.3) = 0.0315 grain/scf

Answers: 213 scf of moist air per pound of fuel oil. 818 ppm SO_2, 34 ppm SO_3, and 0.0315 grain per standard cubic foot of air.

304. BZ air was sampled for total barley dust at 1.8 L/m for 5 hrs, 40 min with a
     37 mm MCE MF with respective pre-sampling and post-sampling weights of
     33.19 mg and 38.94 mg. What was the grain silo filler's eight-hr TWAE expo-
     sure to respirable dust if 85 mass-percent was nonrespirable?

     5 hrs, 40 min = 340 min

     1.8 L/m $\times$ 340 min = 612 L = 0.612 M^3

     38.94 mg − 33.19 mg = 5.75 mg

     5.75 mg/0.612 M^3 = 9.4 mg/M^3 TWAE
     (assuming 340 min represented an 8-hr exposure)

     $$\begin{array}{rcl} 340\ \text{min} \times 9.4\ \text{mg/M}^3 & = & 3196\ \text{mg/M}^3\text{-min} \\ +\underline{140\ \text{min}} \times 0\ \text{mg/M}^3 & = & \underline{0\ \text{mg/M}^3\text{-min}} \\ 480\ \text{min} & & 3196\ \text{mg/M}^3\text{-min} \end{array}$$

     3196 mg/M^3-min/480 min = 6.7 mg/M^3 8-hr TWAE

     15% mass respirable:

     9.4 mg/M^3 $\times$ 0.15 = 1.4 mg/M^3 respirable

     6.7 mg/M^3 $\times$ 0.15 = 1.0 mg/M^3 respirable

     grain dust TLV (oats, wheat, barley) = 4 mg/M^3

     Answers: 9.4 mg/M^3 TWAE to total barley dust assuming the 340-min dust
     exposure represented 8-hr exposure. 6.7 mg/M^3 assuming balance of loader's
     exposure was essentially dust-free. 1.4 mg/M^3 TWAE and 1.0 mg/M^3 TWAE
     as respirable dust for 340-min and 8-hr exposures, respectively (total airborne
     BZ dust, i.e., grain, silica, silicates, pesticides, insect parts, microbes, endo-
     toxins, spores, *ad nauseum*).

305. Air was sampled for HCl gas (mw = 36.45) in 15 mL of impinger solution at
     0.84 L/m for 17 min, 20 sec. HCl collection efficiency was 80%. A chemist
     analyzed 4.7 mcg Cl/mL in the sample and 0.3 mcg/mL in the blank impinger.
     What was the steel pickler's exposure in ppm?

     17 min, 20 sec = 17.33 min   0.84 L/m $\times$ 17.33 min = 14.56 L

     4.7 mcg Cl/mL/0.8 = 5.88 mcg Cl/mL

     5.88 mcg Cl/mL − 0.3 mcg Cl/mL = 5.58 mcg Cl/mL

     5.58 mcg Cl/mL $\times$ 15 mL = 83.7 mcg Cl
     83.7 mcg Cl $\times$ (36.45/35.45) = 86.06 mcg HCl

     $$\text{ppm} = \frac{\dfrac{\text{mcg}}{\text{L}} \times \dfrac{24.45\ \text{L}}{\text{g-mole}}}{\text{molecular weight}} = \frac{\dfrac{86.06\ \text{mcg}}{14.56\ \text{L}} \times 24.45}{35.45} = 4.0\ \text{ppm}$$

     Answer: 4.0 ppm HCl gas. ACGIH TLV for HCl = 5 ppm (C)

306. Determine eight-hr TWAE of a scrap metal processor to Pb dust and fume with exposures of three hrs, 15 min to 17 mcg Pb/M³; 97 min to 565 mcg Pb/M³; and two hrs, 10 min to 46 mcg Pb/M³. The worker wore an approved HEPA dust/fume/mist filter cartridge respirator for 5 1/2 hrs.

concentration (C)	×	time (T)	=	dose (CT) (Haber's Law)
17 mcg Pb/M³	×	195 min	=	3315 mcg/M³-min
565	×	97	=	54805 mcg/M³-min
46	×	130	=	5980 mcg/M³-min
		422		64,100 mcg/M³-min

OSHA Pb PEL = 50 mcg/M³ × 480 min = 24,000 mcg/M³-min

64,100 mcg/M³-min/24,000 mcg/M³-min = 2.67 × PEL

64,100 mcg/M³/480 min = 133.5 mcg Pb/M³ TWAE

If we assume the balance of the 8-hr work shift was "zero" exposure:

(422 min/480 min) × 133.5 mcg/M³ = 117.4 mcg Pb/M³ TWAE

Answers: TWAE = 133.5 mcg Pb/M³, or 2.67 × the PEL. Assuming the balance of the work shift was "zero" exposure: TWAE = 117.4 mcg Pb/M³.

307. A 7000 gallon storage tank in Albuquerque where the barometric pressure is 640 mm Hg contains 2000 gallons of chlorobenzene. Chlorobenzene vapor pressure = 12 mm Hg at 70°F (mw = 112.6). What is the vapor saturation concentration in ppm, %, and mg/M³? What does chlorobenzene smell like? What is its reported odor threshold?

$$\frac{12 \text{ mm Hg}}{640 \text{ mm Hg}} \times 10^6 = 18,750 \text{ ppm} = 1.875\%$$

$$\frac{mg}{M^3} = \frac{ppm \times molecular\ weight}{24.45} = \frac{18,750 \times 112.6}{24.45} = \frac{86,350 \text{ mg}}{M^3}$$

Answers: 18,750 ppm. 1.875%. 86,350 mg chlorobenzene vapor/M³. The odor threshold (detection, not necessarily recognition) for chlorobenzene vapor is 1.3 ppm. Chlorobenzene has been reported to smell like almonds.

308. A chemical plant operator had the following eight-hr TWAEs on Monday: 32 ppm toluene, 19 ppm xylene, and 148 ppm MEK. Their respective TLVs are 50, 100, and 200 ppm. By what percent is the additive exposure limit exceeded?

$$\frac{32 \text{ ppm}}{50 \text{ ppm}} + \frac{19 \text{ ppm}}{100 \text{ ppm}} + \frac{148 \text{ ppm}}{200 \text{ ppm}} = 1.57$$

Answer: 57% in excess of the TLV, or about 3.1 times the action level of 0.5.

309. Air in an empty room (20 × 38 × 12 ft) contains 600 ppm cyclohexene vapor. How long will it take to dilute this to 6 ppm with a 1550 cfm vane-axial exhaust fan? K factor = 3.

$t = ?$ $C_o \rightarrow C$ 600 ppm diluted to 6 ppm

$20 \times 38 \times 12 \text{ ft} = 9120 \text{ ft}^3$ 2.3 volumes for 10% of C_o (or 60 ppm)

another 2.3 volumes for 10% of previous dilution (diluted to 6 ppm), or:

2.3 volumes + 2.3 volumes = 4.6 volumes

$4.6 \text{ volumes} \times 9120 \text{ ft}^3 = 41,952 \text{ ft}^3$

$$41,952 \text{ ft}^3 \times \frac{\text{minutes}}{1550 \text{ ft}^3} = 27.1 \text{ minutes}$$

$K = 3$ $27.1 \text{ min} \times 3 = 81.3 \text{ min}$

$$t = \frac{-\ln\left[\dfrac{C}{C_o}\right]}{\dfrac{Q}{V}} = \frac{-\ln\left[\dfrac{6 \text{ ppm}}{600 \text{ ppm}}\right]}{\dfrac{1550 \text{ cfm}}{9120 \text{ ft}^3}} = \frac{-\ln 0.01}{0.17 \text{ minute}} = \frac{-(-4.605)}{0.17 \text{ minute}} = 27.1 \text{ minutes}$$

Answer: With perfect mixing, the concentration of 600 ppm would dilute to six ppm in 27.1 min. Applying the ventilation imperfect mixing factor of three, dilution time increases to over 81 min.

310. 7.3 μL liquid styrene (mw = 104.2, density = 0.91 g/mL) are evaporated in a 21.6 L glass calibration bottle. What is the styrene vapor concentration in ppm?

7.3 μL   = 0.0073 mL $0.0073 \text{ mL} \times 0.91 \text{ g/mL} = 0.00664 \text{ g} = 6.64 \text{ mg}$
21.6 L   = 0.0216 M³

$$\text{ppm} = \frac{\dfrac{\text{mg}}{\text{M}^3} \times \dfrac{24.45 \text{ L}}{\text{g-mole}}}{\text{molecular weight}} = \frac{\dfrac{6.64 \text{ mg}}{0.0216 \text{ M}^3} \times 24.45}{104.2} = 72.1 \text{ ppm}$$

Answer: 72 ppm.

311. A rotameter was calibrated at 25°C and 760 mm Hg. What is the corrected air flow rate when the rotameter indicates two L/m at 630 mm Hg and 33°C?

$$Q_{actual} = Q_{indicated} \times \sqrt{\frac{P_{cal}}{P_{field}} \times \frac{T_{field}}{T_{cal}}} = \frac{2 \text{ liters}}{\text{minute}} \times \sqrt{\frac{760 \text{ mm Hg}}{630 \text{ mm Hg}} \times \frac{306 \text{ K}}{298 \text{ K}}} = \frac{2.23 \text{ L}}{\text{minute}}$$

Answer: 2.23 L of air/min. *Note:* The square root function must be used with orifice meters (e.g., rotameters and critical orifices). This derives from orifice theory and must be used when an air flow rate orifice functions at pressures and temperatures differing from calibration conditions. In such situations, Charles' and Boyle's laws must not be applied.

312. What is the effective specific gravity of 13,000 ppm of a gas in air when the gas has a specific gravity of 4.6? Will the mixture stratify with the denser gas at floor level?

13,000 ppm = 1.3% (i.e., 98.7% air)

$$
\begin{array}{ll}
0.987 \times 1.0 = & 0.987 \\
+\underline{0.013} \times 4.6 = & +\underline{0.0598} \\
1.000 & 1.0468
\end{array}
$$

Answer: 1.0468 (only 4.7% greater than the density of air). No, never, pointing out the fallacy of placing exhaust air ducts at floor level to capture vapors "heavier than air" or near the ceiling for gases that are "lighter than air." Of course, in those facilities, such as a paint mixing "kitchen," where large volumes of solvents might spill, exhaust air hoods near the floor are often advisable to capture solvent vapors as they evaporate and to promote good ventilation mixing near the areas where a combustible air and vapor mixture could be produced.

313. Analysis of an 866 L MF air sample detected 2667 mcg Zn. How much zinc oxide (ZnO) fume does this represent in the welder's breathing zone? The molecular weights of Zn and O are 65 and 16, respectively.)

ZnO molecular weight = 65 + 16 = 81   81/65 = 1.246
2667 mcg Zn $\times$ 1.246 = 3323 mcg ZnO
3323 mcg ZnO/866 L   = 3.84 mg ZnO/M^3

Answer: 3.84 mg ZnO/M^3.

314. What is air flow rate through an eight-in diameter duct with a transport velocity of 2900 fpm? What capture velocity is expected eight in. in front of duct inlet if there is a wide flange around inlet? Without cross-drafts, what discharge velocity is expected 20 ft from exhaust outlet? What is expected reduction in capture velocity eight in. in front of the exhaust inlet if the wide flange is missing?

area of circle = $\pi r^2 = \pi$ (4 in.)^2 = 50.27 in^2 = 0.349 ft^2

Q = AV = (0.349 ft^2) (2900 fpm) = 1012 cfm

at 8 in.: 290 fpm (one duct diameter)

at 20 ft: 290 fpm (30 duct diameters)

290 fpm $\times$ 0.75 = 218 fpm

Answer: 1012 cfm. 290 fpm at one duct diameter on the exhaust side of the fan. 290 fpm at 30 duct diameters on exhaust air discharge side. Both assume no significant disruptive ventilation cross drafts. Capture velocity on the suction side of the fan increases up to 25% with a wide flange. Therefore, with the absence of a wide flange, the capture velocity one duct diameter in front of the exhaust inlet would be about 75% of the nominal capture velocity with a flange, or 290 fpm − (290 fpm × 0.25) = 218 fpm.

315. Nine detector tube BZ air samples were obtained randomly throughout the work shift with results of 2, 10, 5, 6, 2, 4, 14, 3, and 6 ppm SO_2. What are the worker's arithmetic mean and median exposures? Assuming the air samples are log-normally distributed, what are the standard deviation and 95% confidence range of his or her exposures?

$n = 9$

SO_2, ppm

x	log x	(log x)^2
2	0.301	0.0906
2	0.301	0.0906
3	0.477	0.2275
4	0.602	0.3264
5	0.699	0.4886
6	0.778	0.6053
6	0.778	0.6053
10	1.000	1.0000
14	1.146	1.3133
Σ = 52	Σ = 6.082	Σ = 4.7836

arithmetic mean (the average) = 52 ppm SO_2/9 = 5.8 ppm SO_2

median = antilog 6.082/9 = antilog 0.6758 = 4.5

$$\text{standard deviation} = \text{antilog}\sqrt{\frac{4.7836 - \frac{(6.082)^2}{9}}{9-1}} = \text{antilog}\sqrt{\frac{4.7836 - 4.1101}{8}}$$

antilog $\sqrt{0.0842}$ = antilog 0.2902 = 1.95

Values of the student's t-distribution for 95% confidence range (bilateral test) are:

number of measurements	degrees of freedom	t-value
2	1	12.706
3	2	4.303
4	3	3.182
5	4	2.776
6	5	2.571
7	6	2.447
8	7	2.365

9	8	⇒	**2.306**
10	9		2.262
11	10		2.228
21	20		2.086
31	30		2.042
51	50		2.009
101	100		1.984
501	500		1.965
1001	1000		1.962
∞	∞		1.960

$$95\% \text{ confidence range} = \text{antilog} \left[0.6758 \pm 2.306 \sqrt{\frac{0.0842}{9}} \right] =$$

antilog [0.6758 ± 2.306 (0.0967)] = antilog (0.6758 ± 0.2230)

upper limit = antilog 0.8988 = 7.9 ppm SO_2

lower limit = antilog 0.4528 = 2.8 ppm SO_2

Answers: Arithmetic mean = 5.8 ppm SO_2. Median = 4.5 ppm SO_2. Standard deviation = 1.95. The 95% confidence interval range of the exposures = 2.8 ppm to 7.9 ppm SO_2.

316. A sealed 55-gallon drum containing two gallons of *n*-butylamine has been in an empty room (20 wide × 40 long × 10 ft high) for three weeks. A process operator wearing a full-face airline respirator and protective clothing finishes filling the drum with *n*-butylamine. The air-tight room has no operating ventilation during the filling of the drum. This plant is located in Montana at an elevation where the barometric pressure is 680 mm Hg. Vapor pressure and molecular weight of *n*-butylamine are 82 mm Hg and 73.2, respectively. There is one 1550 cfm exhaust fan in a 20 ft wall with a negative pressure-activated makeup air louver located in the opposite wall. The ventilation mixing factor, K, is estimated at three.

• What is the saturation concentration of *n*-butylamine vapor in the drum in ppm and mg/M^3 before it is filled?
• What is the average *n*-butylamine vapor concentration (in ppm) in the room after the drum has been filled and the bung has been tightened?
• After the full drum is sealed, how long will it take to dilute *n*-butylamine vapor in the room to less than one ppm by operating the exhaust fan?

$$\frac{82 \text{ mm Hg}}{680 \text{ mm Hg}} \times 10^6 = 120,588 \text{ ppm} \cong 12\%$$

Note: LEL = 1.7%   UEL = 9.8%   FP = 10°F   IDLH = 2000 ppm

ACGIH TLV, OSHA PEL, and NIOSH REL = C 5 ppm (C 15 mg/M³) SKIN

$$\frac{mg}{M^3} = \frac{ppm \times molecular\ weight}{24.45} = \frac{120,588 \times 73.2}{24.45} = \frac{361,024\ mg}{M^3}$$

55 gallons – 2 gallons = 53 gallons = 200.6 L
(the volume of saturated air which is displaced into the room)

200.6 L × 361 mg/L = 72,417 mg *n*-butylamine vapor

20 × 40 × 10 ft = 8000 ft³ = 226.6 M³ 72,417 mg/226.3 M³ = 320.0 mg/M³

$$ppm = \frac{\frac{mcg}{L} \times 24.45}{molecular\ weight} = \frac{\frac{320\ mcg}{L} \times 24.45}{73.2} = 106.9\ ppm$$

8000 ft³ × 2.3 room volumes = 18,400 ft³ (yields 10% of 106.9 ppm = 10.7 ppm)

8000 ft³ × 4.6 room volumes = 36,800 ft³ (yields 1% of 106.9 ppm = 1.07 ppm)

36,800 ft³/1550 cfm = 24 min (approximately)

24 min × K (= 3) = 72 min. Therefore, operate the exhaust fan for at least 72 min.

$$t = \frac{-\ln\left[\dfrac{C}{C_o}\right]}{\dfrac{Q}{V}} = \frac{-\ln\left[\dfrac{1\ ppm}{107\ ppm}\right]}{\dfrac{1550\ cfm}{8000\ ft^3}} = \frac{-\ln 0.00935}{0.194\ minute} = \frac{4.673}{0.194\ minute} = 24\ minutes$$

24 min × K (= 3) = 72 min

Answers: 120,588 ppm. 361,024 mg/M³. 107 ppm. > 72 min.

317. An industrial hygienist determines a worker's peak exposure to isopropylamine by drawing the BZ vapor through a small charcoal tube using a 100 mL detector tube pump. Since the pump's orifice samples at a critical rate of 33 mL/min, this method ensures the maximum sampling rate for this size of charcoal tube is not exceeded. TLV for isopropylamine = five ppm. If the true peak BZ concentration is 0.5 ppm, how much vapor will the industrial hygienist collect? Molecular weight of isopropylamine = 59.08.

$$\frac{mg}{M^3} = \frac{mcg}{L} = \frac{ppm \times molecular\ weight}{24.45} = \frac{0.5 \times 59.08}{24.45} = \frac{1.208\ mcg}{L}$$

$$\frac{1.208\ mcg}{L} \times 0.1\ L = 0.1208\ mcg$$

Answer: 0.12 microgram of isopropylamine. Check with the industrial hygiene chemist to ensure that enough vapor has been collected to satisfy the minimum analytically detectable concentration. Otherwise, a larger air sample is required.

318. Calculate the flash point of an aqueous solution containing 75% methyl alcohol by weight. The flash point of 100% methanol is 54°F. Methanol's vapor pressure at this temperature is 62 mm Hg.

Make the calculation based on 100 pounds of solution. The mole fractions of each solution component are needed to apply Raoult's law. Remember that number of moles = mass/molecular weight.

	pounds	molecular weight	moles	mole fraction
methanol	75	32	2.34	0.63
water	25	18	1.39	0.37
			3.73	1.00

Raoult's law is used to calculate the vapor pressure (P_{sat}) of pure methanol based on the partial pressure required to flash, where x = the mole fraction and p = the vapor pressure of the 100% flammable component at its flash point.

$$p = xP_{sat} \quad P_{sat} = \frac{p}{x} = \frac{62 \text{ mm Hg}}{0.63} = 98.4 \text{ mm Hg}$$

Answer: Using a graph of the vapor pressure of methanol vs. temperature, the flash point of the aqueous solution is approximately 67.5°F. Addition of water, if miscible with the flammable solvent, not surprisingly, raises the flash point. See problem 18 for a discussion of applications and the deviations from Raoult's law.

319. The LEL and UEL for a flammable gas are 2.2% and 7.8% (volume/volume), respectively. At the midpoint between the LEL and the UEL ($\cong$ 5%), the explosion pressure is about _____ times greater than an explosion occurring just above the LEL or just below the UEL, respectively.

a. 2
b. 4
c. 10
d. 50
e. 100

Answer: c. This obviously varies from explosive gas to explosive gas, vapor to explosive vapor, and dust to explosive dust, but at the stoichiometric midpoint, the explosion pressure generated is typically an order of magnitude greater than that at the LEL or about an order of magnitude lower than the blast pressure at the UEL.

320. Generally, with excellent mixing of clean dilution ventilation with contaminated air, _____ complete air changes are necessary to ensure that the confined space atmosphere equals the ambient atmosphere concentration.

    a. 2
    b. 5
    c. 10
    d. 20
    e. 53

    Answer: d. That is, the assumption is made the confined space atmosphere is 100% contaminant (i.e., 1,000,000 ppm). After 2.3 air changes, the concentration is reduced to 100,000 ppm. After another 2.3 air changes (4.6 total), the level is reduced to 10,000 ppm. After a total of 6.9 air changes, the concentration is 1000 ppm. After another 2.3 air changes (9.2 total), the level is 100 ppm. After another 2.3 changes (11.5 total), the initial 100% concentration has been reduced to 10 ppm. After 18.4 total air changes, the concentration is reduced to 0.01 ppm, and after 20.7 air changes, the concentration is finally at 0.001 ppm (1 ppb) — a level well below most TLVs and PELs (*bis*-chloromethyl ether and osmium tetroxide are current notable exceptions for gases and vapors).

321. In testing the atmosphere of a confined space for air contaminants, what is the proper sequence (first, second, third)? Flammables and combustibles are to be considered as the same.

    a. toxics, oxygen, flammables
    b. oxygen, flammables, toxics
    c. combustibles, toxics, oxygen
    d. oxygen, toxics, flammables
    e. flammables, oxygen, toxics
    g. toxics, flammables, oxygen
    h. none of the above

    Answer: b. Remember the mnemonic "**OFT**" = **O**xygen, **F**lammables, **T**oxics — and done very **OFT**en.

322. VICI Metronics, Inc. (Santa Clara, CA), a high quality manufacturer of gas and vapor diffusion vials, gives the following example to produce 10 ppm toluene vapor in a 1000 cubic centimeters/min air stream at 30°C:

    a. Calculate the required vapor generation rate:

    $$r = \frac{FC}{K}, \text{ where } K = \frac{24.47 \text{ L/g-mole}}{92.13 = \text{molecular weight}} = 0.266$$

$$r = \frac{(1000 \text{ cc/minute})(10 \text{ ppm})}{0.266} = \frac{37,594 \text{ nanograms}}{\text{minute}}$$

b. Calculate the diffusion rate:

known: molecular weight of toluene = 92.13 (g/mol)
        diffusion vial length = 7.62 cm = L
        T = 30° + 273 K = 303 K
        ρ = 36.7 mm Hg (toluene vapor pressure at 30°C)
        D_0 = 0.0849 cm²/sec (diffusion coefficient at 25°C)
        P = 750 mm Hg (atmospheric pressure)
        A = 0.1963 cm² (5-mm diameter diffusion vial cross-section)

$$r = (1.9 \times 10^4)(303 \text{ K})(0.0849)(92.13)\left[\frac{0.1963}{7.62}\right] \times$$

$$\log\left[\frac{750}{750 - 36.7}\right] = \frac{25,276 \text{ nanograms}}{\text{minute}}$$

c. Length of capillary tube can be shortened to give a higher diffusion rate, e.g.,:

$$L_2 = L_1 \times \frac{r_1}{r_2} = 7.62 \text{ cm} \times \frac{25,276 \text{ ng/min}}{37,594 \text{ ng/min}} = 5.1 \text{ cm}$$

d. To estimate the diffusion rate of a given volatile compound, use the equation:

$$r = 1.90 \times 10^4 \times T \times D_0 \times M \times \left[\frac{A}{L}\right] \times \log\left[\frac{P}{P - \rho}\right]$$

Capillary diffusion tubes can provide a constant source for dynamic gas and vapor calibration systems. They can generate ppb to high ppm concentrations, and their rates are easily calibrated and verified by simple gravimetric procedures. The principle is based on the fact that gases and vapors will diffuse at a steady rate through a capillary tube held at a constant temperature and pressure. The vapor pressure remains constant and serves as the constant driving force for diffusion through a capillary tube. The bore diameter and the diffusion path length then determine the rate for a specific volatile material. Variations in the atmospheric pressure and temperatures and the carrier gas composition will affect the diffusion rate. The actual rate is verified by simply pre- and post-weighing the diffusion tube during the period of use.

323. One gallon of gasoline is accidentally spilled into a 10 × 10 × 10 ft press pit below a large metal stamping machine. The pit does not have a fixed forced mechanical ventilation system. Hours go by. What is the gasoline vapor concentration? What are the hazards? The density of gasoline is 0.75 g/mL. The average molecular weight of gasoline is approximately 73 g/g-mol.

$$\frac{10 \times 10 \times 10 \text{ ft}}{35.3 \text{ ft}^3/\text{M}^3} = 28.32 \text{ M}$$

$$1 \text{ gallon} \times \frac{3785 \text{ mL}}{\text{gallon}} \times \frac{0.75 \text{ g}}{\text{mL}} \times \frac{1000 \text{ mg}}{\text{g}} = 2,838,750 \text{ mg}$$

$$\text{ppm} = \frac{\frac{\text{mg}}{\text{M}^3} \times \frac{24.45 \text{ L}}{\text{g-mole}}}{\text{molecular weight}} = \frac{\frac{2,838,750 \text{ mg}}{28.32 \text{ M}^3} \times 24.45}{73} = 33,572 \text{ ppm}$$

Answers: 33,572 ppm. Note the LEL and UEL for gasoline = 1.4% and 7.6%, respectively. The concentration is in the highly dangerous mid-explosive range. The TLV = 300 ppm (0.03%) with a ceiling of 500 ppm and an IDLH concentration of 5000 ppm (0.5%, or ≅ 1/3 of the LEL). The benzene vapor concentration alone in this vapor mixture is near 300 or more ppm. What if a welder later performs repairs on the press above the pit without taking "hot work" precautions? Does this vapor mixture have a "built-in match", that is, a volatile explosive liquid which, at its saturation concentration, is between the LEL and its UEL?

324. A compressed gas cylinder contains hydrogen at 25°C and at a gauge pressure of 2200 psig. The cylinder's volume is 45 L. What is the mass of hydrogen in this cylinder?

$$P_i V_i = P_f V_f \quad V_f = \frac{P_i V_i}{P_f} = \frac{(2200 \text{ psig} + 14.7 \text{ psia})(45 \text{ L})}{14.7 \text{ psia}} = 6780 \text{ L}$$

$$6780 \text{ L} \times \frac{\text{mole}}{24.45} \times \frac{2 \text{ grams H}_2}{\text{gram-mole}} \times \frac{\text{kg}}{1000 \text{ g}} = 0.555 \text{ kg}$$

Answer: 0.555 kilograms of hydrogen = 1.1 pounds of H_2.

325. Determine the volume that 1.5 mols of diethylsulfide [$(C_2H_5)_2S$] would occupy at 275°C and 12.33 atmospheres. The critical pressure (P_c) = 39.08 atmospheres. The critical temperature (T_c) for $(C_2H_5)_2S$ = 283.8°C.

At this temperature and pressure, this chemical, like many others, does not behave like an ideal gas. Corrections, therefore, are required in the calculations.

$$P_r = \frac{12.33 \text{ atm}}{39.08 \text{ atm}} = 0.316$$

$$T_r = \frac{275°C + 273 \text{ K}}{283.8°C + 273 \text{ K}} = 0.983$$

From tables in physical chemistry handbooks, we can obtain Z from the above correction factors = 0.87. Z is the compressibility factor, an empirical correction factor for the nonideal behavior of real gases.

$$V = \frac{(0.87)(1.50 \text{ moles})(0.0821 \text{ L-atm/mole-K})(548 \text{ K})}{12.33 \text{ atm}} = 4.76 \text{ L}$$

Answer: 4.76 L.

326. The vapor pressure (P_v) can be measured by passing an inert gas over a sample of the material and analyzing the composition of the gaseous mixture. Calculate the vapor pressure of mercury at 23°C and 745 mm Hg if a 50.40 g sample of nitrogen plus mercury vapor contains 0.702 milligram of mercury.

The basic equation is: $\dfrac{P_v}{P_t} = \dfrac{n}{n + n_{inert}}$.

The amount of each gas in the mixture is:

$$n(Hg) = \frac{7.02 \times 10^{-4} \text{ g}}{200.59 \text{ g mole}^{-1}} = 3.50 \times 10^{-6} \text{ mole}$$

$$n(N_2) = \frac{\left[50.40 - \left(7.02 \times 10^{-4}\right)\right] \text{g}}{28 \text{ g mole}^{-1}} = 1.8 \text{ mole}$$

$$P_v = (745 \text{ mm Hg}) \times \frac{3.5 \times 10^{-6}}{\left(3.5 \times 10^{-6}\right) + 1.8} = 1.45 \times 10^{-3} \text{ mm Hg}$$

Answer: 0.00145 mm Hg.

327. A compressed gas cylinder contains 75 L of carbon monoxide at 215 psig and 25°C. If the room atmospheric pressure is 14.4 psi, what mass of CO is vented to the laboratory when the valve is opened?

CO originally in cylinder $= n_1 = \dfrac{PV}{RT} =$

$$\frac{(215 + 14.4)\text{psi}\left[(6895 \text{ Pa})/(1 \text{ psi})\right](75 \text{ L})}{(8314 \text{ L Pa K}^{-1} \text{ mole}^{-1})(298 \text{ K})} = 48.0 \text{ moles}$$

CO remaining at 0 psig $= n_2 = \dfrac{(0 + 14.4) \text{ psi}(6895 \text{ Pa})(75 \text{ L})}{(8314)(298 \text{ K})} = 3.0 \text{ moles}$

$$\left[(48.0 - 3.0) \text{ moles}\right]\left(28 \times 10^{-3} \text{ kg/mole}\right) = 1.3 \text{ kg}$$

Answer: 1,300 g of carbon monoxide gas.

328. A liquid with a molecular weight of 86 evaporates into the air of a workplace from an 8 × 2-ft open surface tank. The vapor pressure of this liquid is 30 mm Hg. The air and liquid temperature are both 25°C. The room air passing over the liquid is 100 ft/min. What is the vapor generation rate?

The vapor generation rate, G (in lb/hr), can be estimated from the EPA equation:

$$G = \frac{13.3792 \, MPA}{T} \times \left[\frac{D_{ab}V_z}{DZ}\right]^{0.5}, \text{ where:}$$

M	= molecular weight (lb/lb-mol)	= 86
P	= vapor pressure (inches of mercury)	= 30 mm Hg = 1.18-in Hg
A	= liquid surface area (ft²)	= 8 × 2 ft = 16 ft²
D_{ab}	= diffusion coefficient	= ?
	(ft²/sec of a through b in air)	
V_z	= air velocity (ft/min)	= 100 ft/min
T	= temperature (K, kelvin)	= 25°C = 298.15 K
ΔZ	= pool or tank length along flow direction	= 8 ft
	(feet)	

D_{ab} can be estimated from the EPA equation:

$$D_{ab} = \frac{4.09 \times 10^{-5}\left(T^{1.9}\right)\left[\frac{1}{29}+\frac{1}{M}\right]^{0.5}\left(M^{-0.33}\right)}{P_t}, \text{ where:}$$

D_{ab}	= diffusion coefficient (cm²/sec)	= ?
T	= temperature (K, kelvin)	= 25°C + 273.15 K = 298.15 K
M	= molecular weight (g/g-mol)	= 86
P_t	= pressure (in atmospheres)	= 30 mm Hg = 0.03947 atm

$$D_{ab} = \frac{4.09 \times 10^{-5}\left(298.15 \text{ K}\right)^{1.9}\left[\frac{1}{29}+\frac{1}{86}\right]^{0.5}\left(86^{-0.33}\right)}{0.03947 \text{ atmosphere}} = 2.57 \text{ cm}^2/\text{second} =$$

$0.00277 \text{ ft}^2/\text{second}$

$$G = \frac{(13.3792)(86 \text{ lb/lb-mole})(1.18" \text{ Hg})(16 \text{ ft}^2)}{298.15 \text{ K}} \times \sqrt{\frac{0.00277 \text{ ft}^2}{\text{second}} \times \frac{100 \text{ ft}}{\text{minute}}}$$

13.56 lb/hr

Answer: Approximately 13-14 pounds of liquid evaporated per hour. From this estimate, one can design industrial hygiene controls (e.g., enclosure, less volatile and/or toxic solvent, local exhaust ventilation, general dilution ventilation, etc.).

329. Assume from the preceding problem that general ventilation (10,000 cfm) is used to dilute the vapors and the ventilation mixing factor = three. What is the contaminant concentration in the workplace?

$$C = \frac{(1.7 \times 10^5)(K)(G)}{M \times Q \times k}, \text{ where:}$$

C = air contaminant concentration (ppm)

K = ambient air temperature (degrees kelvin)

G = vapor generation rate (gm/sec)

M = molecular weight (g/g-mol)

Q = ventilation rate (cfm)

k = ventilation mixing factor (dimensionless, based on subjective judgment)

$$\frac{13.56 \text{ lb}}{\text{hour}} \times \frac{453.6 \text{ grams}}{\text{lb}} \times \frac{\text{hour}}{60 \text{ minutes}} \times \frac{\text{minute}}{60 \text{ seconds}} = \frac{1.709 \text{ grams}}{\text{second}}$$

$$C = \frac{(1.7 \times 10^5)(298.15 \text{ K})(1.709 \text{ g/sec})}{(86 \text{ g/g-mole})(10,000 \text{ cfm})(3)} = 33.6 \text{ ppm}$$

Answer: $\cong$ 34 ppm.

330. The saturation pressure of water vapor in air at 22°C is 19.8 mm Hg. What is the mass concentration of water vapor in air at this temperature when the barometric pressure is 725 mm Hg and the relative humidity is 50%?

$$0.50 \times \left[\frac{19.8 \text{ mm Hg}}{725 \text{ mm Hg}}\right] \times 10^6 = 13,655 \text{ ppm} = 1.37\% \text{ water vapor in air}$$

$$\frac{\text{mg}}{M^3} = \frac{\text{ppm}}{\frac{22.4 \text{ L/g-mole}}{\text{molecular weight}} \times \frac{\text{absolute temperature}}{273.15 \text{ K}} \times \frac{760 \text{ mm Hg}}{725 \text{ mm Hg}}} =$$

$$\frac{13,655 \text{ ppm}}{\frac{22.4}{18} \times \frac{295.15}{273.15} \times \frac{760}{725}} = \frac{9687 \text{ mg H}_2\text{O}}{M^3}$$

Answers: 13,655 ppm = 9687 mg H_2O vapor/M^3.

331. An industrial hygienist using a direct-reading instrument is measuring the mercury vapor in the atmosphere of a chloralkali plant. When the meter indicates a level of 0.04 mg Hg/M^3, there is a release of chlorine gas which, measured with

a detector tube, was about 0.7 ppm Cl_2. At this time, the mercury vapor meter reading fell to 0.01 mg/M^3. What could explain this apparent reduction?

Answer: Mercury vapor detector instruments are only responsive to elemental mercury vapor and not to mercury salts, oxides, or organomercury compounds. Most likely, since chlorine is a strong oxidizer, there was a gas phase reaction of the mercury vapor and chlorine gas to produce mercurous chloride and mercuric chloride salts both of which are not detected by the UV mercury vapor meter. The astute industrial hygienist is constantly vigilant for such possibilities in anomalous results (high or low).

$$\frac{mg}{M^3} = \frac{ppm \times molecular\ weight}{24.45} = \frac{0.7\ ppm \times (2 \times 35.5)}{24.45} = \frac{2.03\ mg\ Cl_2}{M^3}$$

Therefore, the excess chlorine molecules help ensure stoichiometric reactions.

$$Hg^\circ + Cl_2 \rightarrow HgCl_2 \qquad 2\ Hg^\circ + Cl_2 \rightarrow Hg2Cl_2$$

332. A worker inhales 1000 ppm ethyl alcohol vapor continuously throughout his eight-hr work shift. He alleges ethanol intoxication. Is this likely?

molecular weight $CH3CH_2OH = 46.07$ density of EtOH = 0.789 g/mL

$$\frac{mg}{M^3} = \frac{ppm \times molecular\ weight}{24.45\ L/g\text{-}mole} = \frac{1000\ ppm \times 46.07}{24.45} = \frac{1884\ mg}{M^3}$$

During an eight-hr work shift, an average worker inhales approximately 10 cubic meters of air, so:

$$10\ M^3 \times \frac{1884\ mg}{M^3} = 18,840\ mg\ EtOH = 18.84\ g\ EtOH$$

Assuming this worker absorbed 100% of inhaled EtOH (a reasonable assumption since EtOH is readily soluble in mucous membranes and readily absorbed into the systemic circulation):

$$18.84\ g\ EtOH \times \frac{mL}{0.789\ g} = 23.9\ mL\ EtOH$$

This is equivalent to 0.8 ounces of ethyl alcohol, or nearly 1.5 12-ounce bottles of 4.5% ethanol beer — but distributed over eight hrs.

Answer: Although the absorption of ethyl alcohol vapors through the respiratory tract is complete and the absorbed ethanol does not initially pass through hepatic portal circulation, the amount absorbed over eight hrs (during which time there is detoxification at a rate high enough to prevent bioaccumulation to toxic levels) would not result in clinical intoxication. A person on Antabuse therapy, however, could be compromised. For example, exposure of alcoholics taking Antabuse to ethanol vapors is contraindicated because of the possibility of a violent adverse reaction. Furthermore, 1000 ppm of ethanol vapor is, in the author's experience, very irritating to upper respiratory tract mucous membranes. Few workers would willingly tolerate this high vapor exposure for eight hrs.

333. One-micron particles are the optimal size for penetration into, and retention in, the terminal airways, the alveoli. What is the settling velocity of one-micron particles of silica (density of α-quartz, $SiO_2 = 2.65$ g/cM^3) in still air? Will such particles fall from the air and settle to the floor and the ground?

Apply Stoke's law: $v_s = \dfrac{g\,d^2(\rho - \rho_a)}{18\,\eta}$, where:

v_s = the particle settling velocity (in centimeters per second)
g = gravitational attraction of particle (981 centimeter/second)
d = particle diameter (centimeter)
ρ = particle density (g/cM^3)
ρ_a = air density at 25°C (= 0.0017 g/cM^3)
η = coefficient of air viscosity (= 1.828 × 10^{-4} poises) at 25°C

$$n_s = \frac{\left[\dfrac{981\,\text{cm}}{\text{second}}\right](0.0001\,\text{cm})^2\left(2.65\,\text{g/cm}^3 - 0.00117\,\text{g/cm}^3\right)}{(18)\left(1.828 \times 10^{-4}\,\text{poises}\right)} = \frac{0.0079\,\text{cm}}{\text{second}}$$

$$\frac{0.0079\,\text{cm}}{\text{second}} \times \frac{60\,\text{sec}}{\text{min}} \times \frac{60\,\text{min}}{\text{hour}} = \frac{28.4\,\text{cm}}{\text{hour}}$$

Answer: 28.4 cm/hr = 0.93 ft/hr. Normal air currents tend to keep such tiny particles permanently suspended in air until they serve as condensation nuclei for atmospheric moisture or until they flocculate with other particles. Practically speaking, lung-damaging dust does not normally settle out of work place air. These invisible particles remain airborne. *Note:* Particles below one micron in diameter require application of Cunningham's factor to calculate settling rates in still air. This factor, or coefficient, $= C = C'\left[1 + K\left[\dfrac{1}{r}\right]\right]$, where $C' =$ the calculation (as above) using Stoke's law, $K = 0.8$-0.86, and $r =$ particle radius in centimeters. Particles with a radius less than 0.1 micron behave like gas molecules and "settle" according to Brownian motion and the equation:

$$A = \sqrt{\frac{RT}{N}} \times \frac{t}{3\pi\eta r},$$ where $A =$ the distance of motion in time, t; $R =$ the universal gas constant (8.316 × 10^7); $T =$ absolute temperature; $N =$ number of molecules in one mol (6.023 × 10^{23}); $\eta =$ the viscosity of air in poises (1.828 × 10^{-4} at 70°F); and $r =$ the particle radius in centimeters.

334. The smallest particles visible to unaided eyes under the best of viewing conditions (lighting, contrast, color, steadiness, etc.) are about 50 to 100 microns in diameter. Such large particles ("rocks" to industrial hygienists and air pollution engineers) do not penetrate far into the respiratory tract. A 100 micron particle, for example, will not reach the alveolar region of the lungs. Less than 1% of all 100 micron particles penetrate as far as the tracheobronchial region. Close to 50% of the 100 micron particles deposit in the nasopharyngeal region (nostrils and upper throat). The remaining 50% do not deposit and tend to be exhaled. Compare the settling rate of a 100 micron particle in still air to that of a one micron particle as calculated in problem 333.

$$n_s = \frac{\left[\dfrac{981\,cm}{second}\right](0.001\,cm)^2\left(2.65\,g/cm^3 - 0.001117\,g/cm^3\right)}{(18)\left(1.828\times10^{-4}\,poises\right)} = \frac{0.789\,cm}{second}$$

$$\frac{0.789\,cm}{second} \times \frac{60\,sec}{min} \times \frac{60\,min}{hour} = \frac{2840\,cm}{hour}$$

Answer: 2840 cm/hr (= 93 ft/hr = 1.55 ft/min). Even such "large" particles (still with relatively small mass) tend to be buoyant in normal air currents (e.g., 50 fpm often is taken as "still" air). It is easy to see how these larger, nonrespirable particles settling at a rate of 1.55 fpm in quiescent air tend not to settle in an atmosphere having perceptible air motion. That is, imagine a 100 micron particle "settling" at 1.55 fpm being tossed around by a 50 fpm air current — not unlike a cork bobbing in a turbulent sea.

335. In the estimation of a worker's total daily work place exposure to toxic agents (i.e., dose), the industrial hygienist must account for not only inhalation exposures, but those occurring from the percutaneous, or dermal, route as well as from ingestion and intraocularly. In the author's experience, toxicant absorption routes other than inhalation are often not adequately addressed and tend to be discounted. This is no doubt due, more often than not, to difficulty in quantitatively assessing the contribution of these routes of absorption to the overall body burden of a toxicant. With skin-absorbed contaminants, one must not only consider the toxicological legacy of a material to pass through intact skin and its dermal absorption rate (in units, e.g., of $\mu g/cm^2$ of skin surface area/hr), but also the skin contact time, skin surface area contacted, and the type of skin *(e.g., eyelids have the thinnest skin, and the soles of the feet and the palms have the thickest integument)*.

Consider a worker falling into a tank and becoming totally immersed in a liquid at room temperature (the aromatic amine, Methyl Ethyl Death®) which has a dermal absorption rate of 2.3 $\mu g/cm^2/hr$. The elapsed time from when he fell into the tank until he was extracted and totally decontaminated was no more than 17 min. This worker was 6 ft-2 in. tall and weighed 200 pounds. Estimate his total maximum dermal absorption (using worst case assumptions).

The Du Bois formula is used to calculate total body surface area of adult humans:

BSA in M^2 = (weight in kg)$^{0.425}$ × (height in cm)$^{0.725}$ × 0.007184 =

$$\left[200 \text{ lb} \times \frac{\text{kg}}{2.204 \text{ lb}}\right]^{0.425} \times \left[74 \text{ in} \times \frac{2.54 \text{ cm}}{\text{inch}}\right]^{0.725} \times 0.007184 = 2.173 \text{ M}^2$$

2.173 M^2 × (10,000 cm^2/M2) = 21,730 cm^2

$$\frac{2.3 \, \mu g}{\frac{cm^2}{hour}} \times 17 \text{ minutes} \times \frac{\text{hour}}{60 \text{ minutes}} \times 21,730 \text{ cm}^2 = 14,160 \, \mu g \cong 14.1 \text{ mg}$$

Answer: $\cong$ 14.1 milligrams of aromatic amine was the maximum dermally-absorbed dose to which should be added the estimated inhalation and ingestion exposure doses.

336. The velocity pressure in a duct is 0.48 in. of water. The barometric pressure is 640 mm Hg. The air temperature is 190°F. What is the duct air velocity?

P_b = 640 mm Hg = 25.197-in. Hg   460°A + 190°F = 650°A

air density, $D = 1.325 \times \dfrac{P_b}{T_{abs}} = 1.325 \times \dfrac{25.197}{650} = \dfrac{0.05136 \text{ lb}}{ft^3}$

air velocity, fpm $= 1096.2 \times \sqrt{\dfrac{P_v}{D}} = 1096.2 \times \sqrt{\dfrac{0.48" \, H_2O}{0.05136}} = \dfrac{3351 \text{ ft}}{\text{min}}$

Answer: 3351 ft/min. Note how air density decreases, in this case, by almost one-third as the air temperature increases and the barometric pressure decreases.

337. "Standard air" (at 70°F and 29.9-in Hg) has a mass density of:

a. 0.055 lb/ft^3
b. 0.065 lb/ft^3
c. 0.075 lb/ft^3
d. 0.085 lb/ft^3
e. 0.095 lb/ft^3
f. none of the above

Answer: c.

338. An industrial process is generating 17,300 mg of contaminant per hr into a work place atmosphere. It is desired to limit the workers' exposures to this contaminant to no more than 10 mg/M^3 as an eight-hr time-weighted average

exposure by the use of general dilution ventilation. If the ventilation imperfection mixing factor = four, how much ventilation is needed assuming makeup air is contaminant-free?

17,300 mg/hr = 288.3 mg/min

$$V_{reg} = 4\left[\frac{288.3 \text{ mg/minute}}{10 \text{ mg/M}^3}\right] \times \frac{35.3 \text{ ft}^3}{M^3} = \frac{4071 \text{ ft}^3}{\text{minute}}$$

Answer: 4071 cfm. Attempt to reduce the ventilation requirements by reducing generation rate (always the first priority), improving the mixing of fresh air with the contaminated air, and improving industrial hygiene work practices.

339. Direct-reading "real time" random spot breathing zone air sampling of a worker for NO_2 gas during eight hrs gave the following results: 1.3, 0.2, 0.6, 8.1, 15.6, 1.9, 0.5, 0.1, and 27.3 ppm. What is the geometric mean of these test results?

test result	log of test result
1.3 ppm	0.1139
0.2 ppm	−0.6990
0.6 ppm	−0.2218
8.1 ppm	0.9085
15.6 ppm	1.1931
1.9 ppm	0.2788
0.5 ppm	−0.3010
0.1 ppm	−1.0000
27.3 ppm	1.4361
	$\Sigma = 1.7086$

arithmetic average (mean) of Σ of logs = 1.7086 ppm NO_2/9 = 0.19 ppm NO_2

antilog of 0.1898 = geometric mean = $10^{0.19}$ = 1.55 ppm NO_2

Answer: Geometric mean = 1.55 ppm nitrogen dioxide gas.

340. A low-flow personal air sampling pump was calibrated using a 100 mL burette and required 17.3 sec for the soap film bubble to traverse 78 mL. What was the air sampling rate in L/min?

$$\frac{51.3 \text{ seconds}}{60 \text{ seconds/minute}} = 0.855 \text{ minute} \qquad \frac{78 \text{ milliliters}}{1000 \text{ mL/liter}} = 0.078 \text{ liter}$$

$$\text{air flow rate} = \frac{0.078 \text{ liter}}{0.855 \text{ minute}} = \frac{0.091 \text{ L}}{\text{minute}} = \frac{91 \text{ mL}}{\text{minute}}$$

Answer: 0.091 liter of air/min. This low air flow rate does not exceed the recommended maximum rate of 100 mL air/min for small charcoal tubes.

341. 6.6 L of chlorine dioxide gas are inadvertently released into a sealed $12 \times 30 \times 10$ ft laboratory. After completely mixing, what is the ClO_2 gas concentration?

$$(12 \times 30 \times 10 \text{ ft}) \times \left[\frac{28.32 \text{ L}}{\text{ft}^3} \right] = 101.952 \text{ liters}$$

$$\text{ppm} = \frac{6.6 \text{ liters}}{101,952 \text{ liters}} \times 10^6 = 64.7 \text{ ppm } ClO_2$$

Answer: $\cong$ 65 ppm chlorine dioxide gas. Since this is a highly reactive oxidizing gas, the concentration will decay in time as the gas reacts with reducing agents and organic materials in the room. Do not enter this room. without SCBA or until exhaust and dilution ventilation and confirmatory air sampling demonstrate that the ClO_2 gas level is < 0.05 TLV or, better, at nondetectable levels.

342. The air velocity in an eight-in diameter duct is 2500 fpm. What is the new duct velocity as this air passes into a constricted six-in diameter duct opening?

area of circle = πr^2 6 in. Ø area = 0.1963 ft²   8 in. Ø area = 0.3491 ft²

$V_1 \times A_1 = V_2 \times A_2$, where V_1 = air velocity in duct with area A_1 and

V_2 = air velocity in duct with area A_2

$$V_2 = \frac{V_1 \times A_1}{A_2} = \frac{2500 \text{ fpm} \times 0.3491 \text{ ft}^2}{0.1963 \text{ ft}^2} = \frac{4446 \text{ feet}}{\text{minute}}$$

Answer: The new duct velocity is 4446 ft/min.

343. An open-top beaker containing 100 mL of liquid chlorine rests on the edge of a laboratory bench. The Cl_2 gas concentration nine in below the edge of the bench is 10,000 ppm (1%), and the air and gas mixture at this point has a specific gravity of 1.015 (See problem 8). What is the settling rate of this Cl_2 gas and air mixture?

$$V_s = \sqrt{\frac{2g(SG-1)h}{SG}}, \text{ where:}$$

V_s = the settling rate of the gas or vapor mixture, ft/sec
g   = gravity, 32.2 ft/sec²
h   = distance from source, ft
SG   = specific gravity of the air and gas or vapor mixture relative to the specific gravity or density of air, unitless

$$V_s = \sqrt{\frac{2 \times (32.2 \text{ ft/sec}^2) \times (1.015-1) \times 0.75 \text{ ft}}{1.015}} = \frac{0.845 \text{ feet}}{\text{second}}$$

Answer: 0.845 ft per second = 50.7 ft/min, essentially identical to the air velocity taken to be characteristic of "still" air (50 fpm). *Note:* this equation cannot be used for gases or vapors farther than one foot from their emission source.

344. An amateur photographer carelessly poured one liter of 3% sulfuric acid solution (vol./vol.) into an open developing tray in her basement home dark-room. She then leaves not realizing her mistake. The tray contained 500 mL of a 60 mg sodium sulfide/mL solution. Her unventilated darkroom is $5 \times 10 \times 8$ ft. What gas was evolved and in what quantity if 30% of the gas dissolves in the solution? What is the average concentration of the gas in the darkroom's air after mixing?

$$Na_2S + xs\ H_2SO_4 + H_2O \rightarrow Na_2SO_4 + H_2S\uparrow + H_2O$$

Observation of relative amounts of reactants (an excess of H_2SO_4) indicates there will be a stoichiometric reaction: that is, the quantitative conversion of sodium sulfide in a strong acid solution to hydrogen sulfide gas.

$$5 \times 10 \times 8\ \text{ft} = 400\ \text{ft}^3 = 11.33\ \text{M}^3$$

$$500\ \text{mL} \times (60\ \text{mg}\ Na_2S/\text{mL}) = 30,000\ \text{mg}\ Na_2S$$

molecular weight Na_2S = 78.04 g/mol

molecular weight H_2S = 34.08 g/mol

$$\text{moles} = \frac{\text{grams}}{\text{molecular weight}} = \frac{30\ \text{g}\ Na_2S}{78.04} = 0.384\ \text{mole}$$

Therefore, $0.7 \times 0.384 = 0.269$ mol of H_2S was released (i.e., 30% is dissolved).

grams H_2S released = molecular weight of H_2S $\times$ moles released =

$$\frac{34.08\ \text{grams}}{\text{mole}} \times 0.269\ \text{mole} = 9.17\ \text{g}\ H_2S$$

$$\text{ppm} = \frac{\frac{\text{mg}}{\text{M}^3} \times 24.45}{\text{molecular weight}} = \frac{\frac{9170\ \text{mg}\ H_2S}{11.33\ \text{M}^3} \times 24.45}{34.08} = 581\ \text{ppm}\ H_2S$$

Answer: 581 ppm of hydrogen sulfide gas, a potent chemical asphyxiant which produces respiratory paralysis. The gas concentration is far in excess of the 15 ppm ACGIH STEL TLV for H_2S. Large open pans of dangerous liquid chemicals in unventilated darkrooms are an invitation for a toxic gas release incident.

345. A flammable solvent with toxic properties has a molecular weight of 78. What is its vapor density in relation to air?

$$\text{vapor density, a ratio} = \frac{\text{molecular weight of material}}{\text{composite molecular weight of air}} = \frac{78}{29} = 2.69$$

Answer: This vapor is approximately 2.7 times as dense as air. Vapor density ratios are reported at the equilibrium temperature under atmospheric conditions. Unequal or changing conditions can appreciably change the density of any vapor or gas and their mixtures.

346. The standard plumber's torch contains about one pound of propane. An explosion levels a mobile home after repairs were made to a bathroom sink's plumbing. An empty torch was found in the rubble. Litigation ensues with one side saying the propane vapors were the explosion fuel, and the other side alleging that vapors from one ounce of fingernail polish remover (acetone) were the fuel. What appears to be the most plausible cause of the explosion? Assume that the ignition source is unknown, the explosion originated in a closed unventilated bathroom ($6 \times 8 \times 8$ ft), and no one was present at the time of the explosion.

Assume worst case scenarios, starting with propane:

$6 \times 6 \times 8$ ft $= 288$ ft$^3 = 8.16$ M^3 1 lb propane $= 454.6$ g

$$\text{ppm} = \frac{\dfrac{\text{mg}}{\text{M}^3} \times 24.45}{\text{molecular weight}} = \frac{\dfrac{454,600 \text{ mg}}{8.16 \text{ M}^3} \times 24.45}{44}$$

$$= 30,957 \text{ ppm} \cong 3.1\% \text{ gas in air}$$

Since the LEL for propane in air is 2.4%, explosion of propane gas is plausible if almost an entire cylinder was discharged into the air of the unventilated bathroom.

molecular weight and density of acetone, respectively $= 58.1$ and 0.79 g/mL

$$\text{ppm} = \frac{\dfrac{\text{mg}}{\text{M}^3} \times 24.45}{\text{molecular weight}} = \frac{\dfrac{790 \text{ mg/mL}}{8.16 \text{ M}^3} \times \dfrac{29.57 \text{ mL}}{\text{ounce}} \times 24.45}{58.1}$$

$$= 1205 \text{ ppm} \cong 0.12\%$$

Since the concentration of acetone vapor is far below its LEL of 2.5%, the fingernail polish remover solvent cannot be the cause of the explosion.

Answer: In the absence of any other information, the most likely cause of this explosion was the accumulation of propane gas above its LEL and below its UEL in the presence of an ignition source. Moreover, since the plumber said that he could not recall if he closed the torch valve on the fresh cylinder, and the mobile home owner stated in her deposition she never used more than 1/4 ounce of polish remover at a time, the explosion fuel, to a reasonable degree of scientific certainty, was propane gas.

347. An isokinetic stack sample of 98.4 cubic feet of air was collected for sulfuric acid mist in 100 mL of slightly acidic (nonvolatile acid), unbuffered, distilled water in a Greenburg-Smith impinger. If the pH of the solution decreased from an initial pH of 4.3 to pH of 2.1 after sampling, how much sulfuric acid mist was collected? Assume 100% ionization of sulfuric acid (*Note:* See below *).

$$H_2SO_4 \leftrightarrow 2H^+ + SO_4 = 98.4 \text{ ft}^3 \times \frac{28.32 \text{ L}}{\text{ft}^3} = 2786.7 \text{ L} = 2.787 \text{ M}^3$$

A solution of pH 2.1 contains $10^{-2.1}$ mols of hydrogen ion per liter. Similarly, since the initial pH was 4.3, the collection solution hydrogen ion concentration was $10^{-4.3}$ mols H^+ per liter before air sampling began.

$10^{-2.1}$ = 0.00794 mol H^+ /L $10^{-4.3}$ = 0.00005 mol H^+ /L

increase in hydrogen ion = (0.00794 − 0.00005) mols/L = 0.00789 mol H^+ /L

(0.00789 mol H^+ /L) × 0.1 L = 0.000789 mol H^+

Since one mol of H_2SO_4 produces two mols of H^+ :

(0.000789 mol/2) = 0.0003945 mol H_2SO_4 was collected.

molecular weight H_2SO_4 = 98.07 g/mol

(98.07 g/mol) × 0.0003945 mol = 0.03869 g H_2SO_4

$$\frac{38.69 \text{ mg } H_2SO_4}{2.787 \text{ M}^3} = \frac{13.88 \text{ mg } H_2SO_4}{\text{M}^3}$$

* Actually, this result is low since sulfuric acid, as a diprotic acid, ionizes in two stages. The first stage: $H_2SO_4 \leftrightarrow H^+ + HSO_4^-$ is essentially 100% ionization, that of a very strong acid. The second stage: $HSO_4^- \leftrightarrow H^+ + SO_4^-$ is that of a weak acid where the ionization constant is 1.3×10^{-2} . Through the use of the quadratic equation and somewhat exotic calculations, one can determine precisely what the sulfuric acid mist concentration was. Practically, it would be far easier and maybe more accurate to use a specific ion electrode for sulfate ion and then calculate the amount of sulfuric acid by using the ratio of the molecular weights of sulfate ion to sulfuric acid. The total hydrogen ion concentration at equilibrium is the sum of the concentrations due to both stages of ionization. For example, for a 0.01 molar solution of sulfuric acid, the hydrogen ion is 0.0147 molar (0.01 from the first stage of ionization, and 0.0047 from ionization of hydrogen sulfate ion). The pH of the solution would be −log [H^+] = −log (0.01 + 0.0047) = −log 0.0147 = pH 1.83.

Answer: 13.88 milligrams of H_2SO_4 mist per cubic meter of stack exhaust air. If this is an unprotected steel or "white metal" stack and pollution control equipment, severe corrosion is imminent. Perhaps a fiberglass reinforced PVC stack leading to a high efficiency caustic scrubber should be quickly considered.

348. A process releases sulfur dioxide gas at a steady rate into an occupied work area. What is the generation rate of the gas in cubic feet per hr if 88,000 cubic feet of air/min are needed to dilute the SO_2 to two ppm?

$$\frac{88,000 \text{ ft}^3}{\text{minute}} = \left[\frac{\text{generation rate, cfm}}{2 \text{ ppm}}\right] \times 10^6, \text{ or:}$$

$$\text{generation rate, cfm} = \frac{(88,000 \text{ cfm}) \times 2 \text{ ppm}}{10^6} \times \frac{60 \text{ minutes}}{\text{hour}} = \frac{10.56 \text{ ft}^3 \text{ SO}_2}{\text{hour}}$$

Answer: 10.56 cubic feet of sulfur dioxide gas are released per hour. If better control of the leak could be achieved, substantial reduction in dilution air could be considered. Control at the source is always desirable over "capturing the horse once s/he's out of the barn." Good enclosure and confinement certainly beats using a lasso. And since gaseous molecules tend to move around, they will be difficult to capture once they are free.

349. A composite mixture of shredded plastic waste contains 9% chlorine primarily from polyvinyl chloride polymers and copolymers. How much hydrochloric acid gas is released from 100% quantitative, stoichiometric combustion of this PVC waste?

$$RH\text{-}Cl + O_2 \rightarrow RO_2 + H_2O + HCl$$

correct for the conversion of Cl to HCl: $\dfrac{\text{molecular weight HCl}}{\text{molecular weight Cl}} = \dfrac{36.5}{35.5} = 1.028$

Base the calculations on 100 pounds of plastic waste: 100 pounds $\times$ 0.09 $\times$ 1.028 = 9.25 pounds of HCl

Answer: 9.25 pounds of hydrochloric acid gas released from 100% combustion of every 100 pounds of composite plastic waste.

350. The vapor pressure of the very highly volatile diethyl ether is 401 mm Hg at 18°C. What is its vapor pressure at 32°C?

The molar heat of vaporization (ΔH_{vap}) of a chemical is defined as the energy (in kilojoules, normally) required to vaporize one mol of liquid phase of the chemical. For diethyl ether ("ether"), $\Delta H_{vap} = 26.0$ kJ/mol $= 26,000$ joules/mol. Solution requires rearrangement of the Clausius-Clapeyron equation:

$$\ln P = -\frac{\Delta H_{vap}}{RT} + C, \text{ where } C = \text{a constant} \qquad \ln\frac{P_1}{P_2} = \frac{\Delta H_{vap}}{R} \times \frac{T_1 - T_2}{T_1 \times T_2}$$

$P_1 = 401 \text{ mm Hg} \qquad P_2 = ?$
$T_1 = 18°C = 291 \text{ K} \quad T_2 = 32°C = 305 \text{ K}$

$$\ln\frac{401 \text{ mm Hg}}{P_2} = \frac{26{,}000 \text{ Joules/mole}}{8.314 \text{ Joules/K-mole}} \times \left[\frac{291 \text{ K} - 305 \text{ K}}{(291 \text{ K})(305 \text{ K})}\right]$$

taking the antilog of both sides: $\dfrac{401 \text{ mm Hg}}{P_2} = 0.6106$

$P_2 = 657 \text{ mm Hg}$

Answer: The vapor pressure of diethyl ether at 32°C is 657 mm Hg, close to its boiling point of 94°F at 760 mm Hg (sea level).

351. The half-life of a chemical in air due to atmospheric oxidation is 4.7 hrs. If the initial air concentration of this chemical was 19.8 mg/M^3, how much is left in the air after 12 half-lives?

We can generalize the fraction of air contaminant left after n half-lives as $\left[\dfrac{1}{2}\right]^n$.

$$\frac{19.8 \text{ mg}}{M^3} \times \left[\frac{1}{2}\right]^{12} = \frac{19.8 \text{ mg}}{M^3} \times 0.000244 = \frac{0.00483 \text{ mg}}{M^3}$$

Answer: 4.83 micrograms per cubic meter after (4.7 hrs $\times$ 12 =) 56.4 hrs.

352. What is the half-life for an unstable air contaminant which decays by following first order kinetics if the initial gas concentration is 367 ppm and the concentration after 39.3 hrs is 1.6 ppm? Assume there is no further gas generation when timing begins and that the loss is due entirely to chemical change and not due to loss by ventilation or other means.

This problem is similar to problem 351.

367 ppm, $C_o \rightarrow$ 1.6 ppm, C, after 39.3 hrs

$$(367 \text{ ppm})\left[\frac{1}{2}\right]^n = 1.6 \text{ ppm} \qquad (0.5)^n = \frac{1.6 \text{ ppm}}{367 \text{ ppm}} = 0.00436$$

$$n = \frac{\log 0.00436}{\log 0.5} = \frac{-2.3605}{-0.301} = 7.84 \text{ half-lives} \qquad \frac{39.3 \text{ hours}}{7.84 \text{ half-lives}} = \frac{5.01 \text{ hours}}{\text{half-life}}$$

Answer: $T_{1/2} = 5.0$ hrs or, after one hr, 184 ppm contaminant remains.

353. The OSHA PEL for lead (on a mass to volume basis) is 50 micrograms of lead per cubic meter of air. What is the PEL for lead if it was expressed on a mass to mass basis?

Use the mass of lead to the mass of air containing the lead aerosol.

$$\frac{50 \text{ mcg Pb}}{M^3} \times \frac{M^3}{35.3 \text{ ft}^3} \times \frac{\text{ft}^3}{0.075 \text{ lb}} \times \frac{\text{lb}}{454 \text{ g}} \times \frac{\text{g}}{10^6 \text{ mcg}} = \frac{50 \text{ mcg Pb}}{1202 \times 10^6 \text{ mcg air}} =$$

$$\frac{1 \text{ mcg Pb}}{2.4 \times 10^7 \text{ mcg air}} = \frac{1 \text{ Pb}}{2.4 \times 10^7 \text{ air}} = 4.2 \times 10^{-8}$$

Answer: 1 part of lead by weight in approximately 24,000,000 parts of air by weight. This equals 42 parts of lead per billion parts of air, both by weight.

354. Determine internal pressure exerted by 498 g of sulfur pentafluoride gas at 95°F in a 6.4 L steel vessel. Molecular weight of $S_2F_{10} = 254.1$. IDLH = one ppm.

$$\frac{498 \text{ grams}}{254.1 \text{ grams/mole}} = 1.96 \text{ moles} \quad 95°F = 35°C$$

$$P = \frac{nRT}{V} = \frac{(1.96 \text{ moles})(0.0821 \text{ L-atm / K-mole})(35°C + 273 \text{ K})}{6.4 \text{ L}} = 7.74 \text{ atm}$$

Answer: 7.74 atmospheres.

355. Air bags in automobiles are rapidly detonated during a crash from decomposition of sodium azide, NaN_3, according to the equation:

$$NaN_3(\text{solid}) \xrightarrow{\text{BOOM!}} 2 \text{ Na (solid aerosol)} \uparrow + 3N_2(g) \uparrow$$

The released nitrogen gas rapidly inflates the "air" bag. Determine the volume of N_2 gas generated at 30°C and 640 mm Hg upon the decomposition of 70 g of NaN_3 assuming quantitative stoichiometry.

Two mols of sodium azide ⇒ three mols of nitrogen.   30°C + 273 K = 303 K

$$\text{moles N}_2 \text{ gas} = 70 \text{ g NaN}_3 \times \frac{1 \text{ mole NaN}_3}{65.02 \text{ g NaN}_3} \times \frac{3 \text{ moles N}_2}{2 \text{ moles NaN}_3}$$

$$= 1.615 \text{ moles of N}_2$$

$$V - \frac{nRT}{P} = \frac{(1.615 \text{ moles})(0.0821 \text{ L-atm/K-mole})(303 \text{ K})}{\frac{640 \text{ mm Hg}}{760 \text{ mm Hg}}} = 47.7 \text{ L N}_2$$

Answer: 47.7 L of nitrogen gas. The literature reports that 100 g of sodium azide produces 1.64 mols of gas. Since 70 g sodium azide theoretically generates 1.62 mols of nitrogen gas, 100 g should release 2.31 mols of N_2. However, since only 1.64 mols of nitrogen are generated, the reaction is about 71% quantitative. That is, it appears that there is about 29% undetonated or partial, nongaseous explosion reaction products (NaN_3 aerosol $\Rightarrow$ Na $\Rightarrow$ NaOH $\Rightarrow Na_2CO_3 \Rightarrow NaHCO_3$).

356. In the previous problem, assume the released sodium quickly reacts with moisture in the air to form a sodium hydroxide aerosol. Assuming dispersion in two cubic meters of the car's interior atmosphere, calculate the average concentration of NaOH in the aerosol phase. Any unreacted sodium aerosol in contact with moist mucous membranes will also produce caustic, irritating NaOH. In a relative risk analysis, balance the inhalation and ocular exposure to this caustic aerosol with the safety benefits of automobile air bags.

One mol of sodium azide produces one mol of sodium:

$$\frac{70 \text{ g NaN}_3}{65.02 \text{ g NaN}_3/\text{mole}} = 1.077 \text{ mole NaN}_3 \Rightarrow 1.077 \text{ mole Na}$$

$$2 \text{ Na} + 2 \text{ H}_2\text{O} \rightarrow 2 \text{ NaOH} + \text{H}_2\uparrow$$

Therefore, a maximum of 1.077 mols of sodium hydroxide are produced.

$$\frac{40 \text{ grams NaOH}}{\text{mole}} \times 1.077 \text{ moles} = 43.08 \text{ grams of NaOH}$$

$$\frac{43.08 \text{ g NaOH}}{2 \text{ M}^3} \times \frac{1000 \text{ mg}}{\text{gram}} = \frac{21,540 \text{ mg NaOH}}{\text{M}^3}$$

Answer: 21,540 mg NaOH/M^3 — momentarily intensely irritating, but one's life might have been saved by deployment of the "air" bag. Regardless, sodium and sodium hydroxide particulate aerosol in the eyes, especially, and in contact with moist mucous respiratory tract membranes could cause significant injuries if not immediately flushed with water. Those wearing contact lenses are particularly susceptible to corrosive chemical burns of the conjunctivae, corneas, and eyelids.

357. Consider the hydrogen gas generated in the previous problem. Could this, under the worst of circumstances, result in an explosive atmosphere? Assume that the driver was smoking as the air bag detonated and that his or her cigarette was not extinguished (or that there were other sources of ignition). Further assume that the released hydrogen gas is mixed with ambient air in a 500 L volume.

Each mole of Na produced yields 0.5 mol H_2. Regardless, it is unlikely sodium aerosol would react completely, in even the most humid atmosphere, to provide a stoichiometric, rapid conversion to NaOH and hydrogen gas. Assuming the direst:

1.077 mol sodium $\Rightarrow$ 0.539 mol hydrogen gas

$$\frac{2 \text{ grams } H_2}{\text{gram-mole}} \times 0.539 \text{ mole} = 1.078 \text{ grams of hydrogen gas}$$

$$ppm = \frac{\frac{mg}{M^3} \times 24.45}{\text{molecular weight}} = \frac{\frac{1078 \text{ mg}}{0.500 \text{ M}^3} \times 24.45}{2} = 26,357 \text{ ppm}$$

Answer: 2.64% hydrogen. LEL for $H_2 = 4\%$. UEL for $H_2 = 75\%$. The hydrogen gas concentration, during one of the worst scenarios, is below the LEL and far below stoichiometric midpoint for producing a maximum explosion blast pressure. A hydrogen gas explosion does not appear likely under these conditions.

358. A 10,000 gallon air-tight carbon steel tank is tightly sealed when the atmospheric pressure is 760 mm Hg. The tank's interior is not protected by paint, an oil film, or any other coating. The tank contains only air and atmospheric moisture. After 10 weeks, the interior pressure of the tank is 630 mm Hg. Assume the temperature remains constant at 25°C. What is happening?

Atmospheric oxygen is reacting with iron in the steel to form iron oxides. As O_2 is consumed, its partial pressure is reduced, and the overall pressure in the tank is reduced. Rust forms on the steel surface.

3 Fe (s) + 2 O_2 (g) $\rightarrow$ Fe_3O_4 (s)
4 Fe (s) + 3 O_2 (g) $\rightarrow$ 2 Fe_2O_3 (s)

There is a concurrent percentage increase in inert gases (nitrogen and argon) in the atmosphere as oxygen is consumed. The amount of rust formed would not appreciably change the gas volume of this tank and may be disregarded in the following calculations.

The number of moles of oxygen consumed is related to the drop in pressure, or:

initial tank pressure = (760 mm Hg/760 mm Hg) = 1.000 atmosphere

final tank pressure = (630 mm Hg/760 mm Hg) = 0.829 atmosphere

1.000 atmosphere − 0.829 atmosphere = 0.171 atmosphere which corresponds to the chemical consumption of O_2 (oxidation of iron).   25°C + 273 K = 298 K

10,000 gallons = 37,854 L

$$n = \frac{PV}{RT} = \frac{(0.171\,\text{atm})(37{,}854\,\text{L})}{(0.0821\,\text{L-atm}/\text{K-mole})(298\,\text{K})} = 264.6 \text{ moles of } O_2 \text{ consumed}$$

$$264.4 \text{ moles of } O_2 \times \frac{32 \text{ grams } O_2}{\text{mole}} = 8461 \text{ grams of } O_2 \text{ were consumed.}$$

The tank originally contained 10,000 gallons of air at 21% oxygen by volume. The original molar concentration of oxygen in the tank was:

$$n = \frac{PV}{RT} = \frac{(0.21\,\text{atmosphere})(37{,}854\,\text{L})}{(0.0821\,\text{L-atm}/\text{K-mole})(298\,\text{K})} = 324.9 \text{ moles of } O_2$$

The oxygen in the tank was reduced to $\left[1 - \left[\dfrac{264.4 \text{ moles}}{324.9 \text{ moles}}\right]\right] \times 100 = 18.6\%$

of the initial oxygen concentration.

Answer: The tank atmosphere has become substantially oxygen deficient, i.e., 21% O_2 × 0.186 = 3.9% O_2 by volume. Inhalation of this atmosphere would cause immediate collapse and death within min if the entrant was not rescued and given immediate CPR and EMS care. Carefully follow OSHA confined space entry procedures (29 *CFR* 1910.146).

359. A diver wearing a self-contained under water breathing apparatus (SCUBA) is 20 ft deep in sea water. If, without breathing, he quickly rose to the water's surface, what would happen to the air in his lungs?

Sea water is denser than fresh water: 1.03 g/mL *vs.* 1.00 g/mL. Therefore, the pressure exerted by a 33 ft sea water column is equivalent to one atmosphere pressure. Since pressure increases with increasing depth, at 66 ft the pressure of the water is equivalent to two atmospheres, 99 ft ≈ three atmospheres, etc. As ascent started at 20 ft below the surface, the total decrease in the pressure

for this depth is $\left[\dfrac{20 \text{ feet}}{33 \text{ feet}}\right] \times 1$ atmosphere = 0.606 atmosphere. When the

diver reaches the surface, the air volume trapped in his lungs would have

increased by a factor of $\dfrac{(1 + 0.606)\,\text{atm}}{1\,\text{atm}} = 1.606$ times. This rapid expansion

can fatally rupture the lung's delicate membranes. Development of air embolism is another serious possibility where the expanded air in the lungs is squeezed into the pulmonary capillaries.

Answer: Air in the diver's lungs would expand, without his breathing, 1.6 times with a possible air embolism, pulmonary rupture, coma, and death.

360. A diver will descend to a water depth where the total pressure is equivalent to two atmospheres. What should the oxygen content of his SCUBA air be?

When P_T is the total gas pressure, the oxygen partial pressure, P_{O_2}, is given by:

$$P_{O_2} = X_{O_2} = \frac{n_{O_2}}{n_{O_2} + n_{N_2}} \times P_T .$$

n_{N_2} = partial pressure due to inert gases (nitrogen, argon). However, since gas volume is directly proportional to the number of moles of gas present (at constant temperature and pressure):

$$P_{O_2} = \frac{V_{O_2}}{V_{O_2} + V_{N_2}} \times P_T .$$

Thus, composition of air is 21% oxygen gas by volume and 79% inert gases by volume. When a diver is submerged, the composition of the air must be changed. At a depth equivalent to 2.0 atmospheres, oxygen content of the air should be reduced to 10.5% by volume to maintain the same partial pressure of 0.21 atmosphere:

$$P_{O_2} = 0.21 \text{ atmosphere} = \frac{V_{O_2}}{V_{O_2} + V_{N_2}} \times 2.0 \text{ atmospheres, or:}$$

$$\frac{V_{O_2}}{V_{O_2} + V_{N_2}} = \frac{0.21 \text{ atmosphere}}{2.0 \text{ atmosphere}} = 0.105, \text{ or } 10.5\% \text{ by volume}$$

Answer: 10.5% oxygen gas by volume.

361. A stainless steel tank leaks formaldehyde gas at the rate of 0.3 mL/hr when the internal gas pressure is three atmospheres. What would the leakage effusion rate be if the tank contained vinyl chloride gas?

$$\frac{\text{HCHO leakage rate}}{\text{VC leakage rate}} = \sqrt{\frac{\text{VC molecular weight}}{\text{HCHO molecular weight}}} = \sqrt{\frac{62.49 \text{ grams/mole}}{30.03 \text{ grams/mole}}} =$$

$$\sqrt{2.0809} = 1.443$$

$$\text{VC leakage rate} = \frac{\text{HCHO leakage rate}}{1.443} = \frac{0.30 \text{ mL/hour}}{1.443} = \frac{0.208 \text{ mL}}{\text{hour}}$$

Answer: 0.208 mL of vinyl chloride gas effuses per hr.

362. Dinitrogen pentoxide, N_2O_5, decomposes according to first-order reaction kinetics. At 50°C, the rate constant is approximately 0.00054/second. This linear reaction dissociation of one mol of nasty gas into two mols of another evil gas is:

$$2 \ N_2O_5 \rightarrow 4 \ NO_2 + O_2$$

What is the N_2O_5 concentration after 17.3 min if the initial gas concentration was 3.9 ppm? How long does it require for the initial gas concentration to decay to 0.4 ppm? How long does it take to convert 50% of the initial gas concentration?

$$\ln \frac{conc_{initial}}{conc_{final}} = kt = \ln \frac{3.9 \text{ ppm}}{conc_{final}} = (0.00054/second) \left[17.3 \text{ minutes} \times \frac{60 \text{ seconds}}{minute} \right]$$

$$\ln \frac{3.9 \text{ ppm}}{conc_{final}} = 0.5605 \qquad \frac{3.9 \text{ ppm}}{conc_{final}} = e^{0.5605} = 1.75$$

$$conc_{final} = \frac{3.9 \text{ ppm}}{1.75} = 2.23 \text{ ppm } N_2O_5$$

$$\ln \frac{3.9 \text{ ppm}}{0.4 \text{ ppm}} = (0.00054/second) \, t \qquad \ln 9.75 = (0.00054/second) \, t$$

$$t = \frac{\ln 9.75}{0.00054/second} = \frac{2.277}{0.00054/second} = 4217 \text{ seconds} = 70.3 \text{ minutes}$$

$$t = \frac{1}{k} \ln \frac{conc_{initial}}{conc_{final}} = \frac{1}{0.00054/second} \times \ln \frac{1}{0.5} = 1852 \text{ seconds} \times \ln 2 = 1852 \times$$

$$0.693 = 1283 \text{ seconds} = 21.4 \text{ minutes}$$

Answers: 2.23 ppm. 70.3 min. 21.4 min.

363. An industrial hygienist and a safety engineer collaborate to determine the bursting temperature of a steel gas vessel. If this vessel contains a fixed mass of gas at 25°C at three atmospheres, and it can withstand a pressure of 20 atmospheres, what is the maximum temperature to which this vessel can be increased before it bursts?

Apply Gay-Lussac's law: at constant volume, the pressure of a mass of gas varies directly with the absolute temperature:

$$\frac{P_i}{T_i} = \frac{P_f}{T_f}, \text{ or } T_f = \frac{P_f T_i}{P_i} = \frac{(20 \text{ atm})(273 \text{ K} + 25°C)}{3 \text{ atm}} = \frac{(20 \text{ atm})(298 \text{ K})}{3 \text{ atm}} = 1987 \text{ K}$$

Answer: 1987 kelvin, well above steel's melting point of approximately 1380°C (1653 kelvin). However, a safety factor must be applied: either reduce maximum temperature rating and/or increase the bursting strength of the gas vessel.

364. Some commercial drain cleaners contain aluminum powder and sodium hydroxide (NaOH). The following reaction occurs when a dry mixture of these two powders is poured into the water of a greasy, clogged drain:

$$2 \text{ NaOH} + 2 \text{ Al} + 6 \text{ H}_2\text{O} \rightarrow 2 \text{ NaAl(OH)}_4 + 3 \text{ H}_2\uparrow$$

The generated heat aids in melting any grease. NaOH saponifies grease and fats. As hydrogen is evolved, the gas bubbles stir up the solids which plug the drain. If four g of aluminum are added to a drain containing an excess of NaOH, could sufficient hydrogen gas be evolved into a one-cubic foot volume above the drain to cause an explosion if there was a source of ignition? Atomic weight of aluminum = 26.98 g/mol. Assume the temperature above the drain is 25°C. Disregard any solubility of hydrogen gas into the standing water within the clogged drain.

One mol of aluminum $\Rightarrow$ 1.5 mols of hydrogen gas.

$$\frac{4.0 \text{ grams Al}}{26.98 \text{ grams/mole}} = 0.148 \text{ mole of Al}$$

0.148 mol Al $\times$ 1.5 = 0.222 mol of hydrogen gas

$$0.222 \text{ mole} \times \frac{2 \text{ grams hydrogen}}{\text{mole}} = 0.444 \text{ gram H}_2 \quad \text{one ft}^3 = 28.32 \text{ L}$$

$$\text{ppm} = \frac{\frac{\text{mg}}{\text{M}^3} \times 24.45}{\text{molecular weight}} = \frac{\frac{444 \text{ mg}}{0.02832 \text{ M}^3} \times 24.45}{2} = 191,663 \text{ ppm} \cong 19.2\%$$

Answer: Under this nearly worst case scenario, an explosion would occur since the lower explosive limit of hydrogen gas is 4%. The UEL is 75%. Suppliers of these drain cleaners recommend immediately placing a container over the drain after the chemicals are added to contain any eruption of corrosive, hot alkali into one's face and eyes. This is good advice, however chemical splash goggles and rubber gloves and a face shield must be worn before chemicals are added. Since an explosion could occur, ignition sources must be prohibited, and ventilation is necessary to dilute any hydrogen gas to at least 20% below its LEL. Furthermore, one might not be able to place a container over the drain quickly enough since the chemicals could react instantaneously. Of all the chemicals used to clean drains, concentrated sulfuric acid, in the author's opinion, must never be used because of its "wild card" behavior and incredibly high heat of dilution. Mechanical aids such as a plunger ("plumber's helper"), drain snakes, boiling water, and compressed air should always be tried before using corrosive chemicals. Reliance upon plumbers might be more prudent than using concentrated H_2SO_4, HCl, NaOH, and KOH. The average homeowner does not have sufficient chemical sophistication to use such highly hazardous agents.

365. The ozone gas molecules present in the stratosphere absorb much of the harmful radiation emitted from the sun. A typical temperature and pressure of ozone in the stratosphere are 0.001 atmosphere and 250 kelvin. How many ozone molecules are in each liter of stratospheric air under these upper atmospheric conditions?

$$PV = nRT$$

$$n - \frac{PV}{RT} = \frac{(0.001\,\text{atm})(1.0\,\text{L})}{(0.0821\,\text{L-atm/mole-K})(250\,\text{K})} = 0.0000487\,\text{mole O}_3$$

$$0.0000487\,\text{mole O}_3 \times \frac{6 \times 10^{23}\,\text{molecules O}_3}{\text{mole}} = 2.9 \times 10^{19}\,\text{molecules O}_3/\text{L}$$

Answer: 2.9×10^{19} molecules of ozone gas per liter.

366. What is the partial pressure of dioxane (in atmospheres) at 12.0 ppm vapor in air by volume if the total barometric pressure is 730 mm Hg and the temperature is 16°C?

$$P_{\text{dioxane}} = X_{\text{dioxane}} \times P_{\text{total}} = \frac{12}{10^6} \times 730\,\text{mm Hg} \times \frac{1\,\text{atm}}{760\,\text{mm Hg}} = 1.15 \times 10^{-5}\,\text{atm}$$

Answer: 0.0000115 atmosphere partial pressure from dioxane vapor. The air temperature does not enter into the calculations. Refer to Dalton's law of partial pressures.

367. If an average adult inhales 320 mL of dioxane vapor at the concentration in the previous problem with each breath, how many molecules of dioxane are inhaled?

$$PV = nRT \quad n = \frac{PV}{RT} = \frac{(1.15 \times 10^{-5}\,\text{atm})(0.32\,\text{L})}{(0.0821\,\text{L-atm/mole-K})(289\,\text{K})} = 1.55 \times 10^{-7}\,\text{mole}$$

$$1.55 \times 10^{-7}\,\text{mole} \times \frac{6 \times 10^{23}\,\text{molecules}}{\text{mole}} = 9.3 \times 10^{16}\,\text{molecules of dioxane vapor}$$

Answer: 9.3×10^{16} molecules of dioxane vapor per inhalation.

368. A 55-gallon drum is splash filled with a volatile solvent. For each gallon of solvent added to the drum, one gallon of air that is approximately saturated with solvent vapors will be displaced. To reduce solvent vapor emissions, a dip pipe extending to within two in of the bottom of the drum is used for filling. What percent of vapor saturation can be expected?

a. 130–150% — Slower filling produces supersaturation, i.e., vapor plus mist.
b. 100% — There is no difference in vapor emissions over splash filling.
c. 80%
d. 50%
e. 30%
f. 10%

Answer: d. Several empirical studies have demonstrated reductions of vapor emissions by approximately 50% when fill tubes are used in place of splash filling of volatile solvents.

369. A breathing zone air sample was obtained from a worker exposed to the dust of barium fluoride, BaF_2, a fluxing agent. Air was sampled at an average rate of 1.64 L/m for 473 min out of the worker's eight-hr work shift. The filter contained 593 micrograms of barium. If this worker had no other exposures to fluoride dust, what were his time-weighted average exposures to barium and fluoride that day? Molecular weights of BaF_2 and Ba = 175.33 and 137.33, respectively.

473 min ≅ 8 hrs (98.5% of the full work shift). Reduce exposure calculation by 1.5%.

$$\frac{1.64 \text{ liters}}{\text{minute}} \times 473 \text{ minutes} = 775.7 \text{ liters of air were sampled}$$

$$\frac{593 \text{ mcg Ba}}{775.7 \text{ L}} = \frac{0.764 \text{ mg Ba}}{M^3} \qquad \frac{\text{mol. wt. BaF}_2}{\text{mol. wt. Ba}} = \frac{175.33}{137.33} = 1.277$$

593 mcg Ba × 1.277 = 757.3 mcg BaF_2

$$757.3 \text{ mcg BaF}_2 - 593 \text{ mcg Ba} = 164.3 \text{ mcg F} \quad \frac{164.3 \text{ mcg F}}{775.7 \text{ L}} = \frac{0.212 \text{ mg F}}{M^3}$$

Answers: 0.764 mg Ba/M³ and 0.212 mg F/M³. Clearly, the exposure exceeds the TLV of 0.5 mg/M³ for barium (soluble compounds). The TLV of 2.5 mg/M³ for soluble fluoride compounds is not exceeded. The toxic effects of both are most likely not additive. Barium fluoride is only sparingly soluble in water (0.12 g/100 cc at 25°C), so whether there is truly an overexposure would be a tough judgment call. Prudence would dictate that, since little is known about the chronic toxicity of barium fluoride, better industrial hygiene is needed to reduce his exposure to well below 0.5 mg Ba/M³.

370. Use the thermal mass balance calculation method to estimate outdoor air quantity when the temperature of the HVAC system return air is 76°F, the temperature of the mixed air is 72°F, and the temperature of the outdoor air is 42°F.

$$\text{Outdoor air (percent)} = \frac{T_{\text{return air}} - T_{\text{mixed air}}}{T_{\text{return air}} - T_{\text{outdoor air}}} \times 100 = \frac{76°F - 72°F}{76°F - 42°F} \times 100 = 11.8\%$$

Answer: 11.8% outdoor air.

371. Use the carbon dioxide measurement calculation method to estimate the outdoor air quantity when the gas concentrations of CO_2 are 440 ppm in the mixed air, 490 ppm in the return air, and 325 ppm in the outdoor air.

$$\text{Outside air (percent)} = \frac{C_S - C_R}{C_O - C_R} \times 100 = \frac{440 \text{ ppm} - 490 \text{ ppm}}{325 \text{ ppm} - 490 \text{ ppm}} \times 100$$

$$= \frac{-50 \text{ ppm}}{-165 \text{ ppm}} \times 100 = 30.3\%$$

Answer: 30.3% outside air.

372. What is the minimum free area for introducing outside air into a combustion device per 2000 BTU input?

a. one square inch
b. two square inches
c. five square inches
d. 10 square inches
e. 27 square inches

Answer: a. One square inch.

373. Painters sometimes, perhaps foolishly, mix a liquid insecticide mixture into paint which they apply on walls, ceilings, and trim. Presumably, any insects which later contact the dry paint absorb the pesticide and subsequently die. Chlorpyrifos has been used for this purpose. The TLV for this organophosphate insecticide, which has a "Skin" notation by the ACGIH TLV Committee, is 0.2 mg/M³. The molecular weight of Chlorpyrifos is 350.57, and its vapor pressure is 1.87×10^{-5} mm Hg at 25°C. A painter added three liquid ounces of an 11% mixture (weight/volume) of Chlorpyrifos to one gallon of paint and painted the four walls and the ceiling of an infant's bedroom. The wall dimensions of the room were 8×10 ft. If we assume the bedroom had no ventilation after the paint dried, what could have been the maximum saturation concentration of Chlorpyrifos vapor? Was there sufficient Chlorpyrifos in the paint to saturate the air with vapor? If the painter sprayed this paint, instead of brushing and rolling, and the average total mist concentration in his breathing zone was one mL of paint/M³ for the one hr it required to paint the room, what was his eight-hr time-weighted average exposure to Chlorpyrifos mist? Assume that the paint was completely used to paint this bedroom; that is, one

gallon of paint typically covers about 400 square feet { in this case: [4 (8 ×
10 ft)] + (10 × 10 ft) = 420 ft²] }. Assume that during paint spraying one
milliliter of paint existed as a mist aerosol in each cubic meter of his or her
breathing zone air, and the density of the paint was 1.4 g/mL.

$$\frac{1.87 \times 10^{-5} \text{ mm Hg}}{760 \text{ mm Hg}} \times 10^6 = 0.0246 \text{ ppm Chlorpyrifos vapor at saturation}$$

$$mg/M^3 = \frac{ppm \times molecular \ weight}{24.45 \text{ L/g-mole}} = \frac{0.0246 \text{ ppm} \times 350.57 \text{ g/mole}}{24.45 \text{ L/g-mole}}$$

$$= 0.35 \text{ mg/M}^3$$

This concentration of Chlorpyrifos vapor did not exceed the Threshold Limit
Value when calculated as an eight-hr time-weighted average exposure; however, since the painter was in the room for only one hr, and it is highly unlikely
the Chlorpyrifos would yield a vapor saturation concentration in this brief
time, he/she most probably, was not excessively exposed to vapor (if he/she
had no skin contact).

$$\frac{11 \text{ g}}{100 \text{ mL}} \times 3 \text{ ounces} \times \frac{29.57 \text{ mL}}{\text{ounce}} = 9.76 \text{ grams of Chlorpyrifos per gallon of}$$
paint

$$\text{bedroom volume} = 10' \times 10' \times 8' \times \frac{M^3}{35.3 \text{ ft}^3} = 22.67 \text{ M}^3$$

$$22.67 \text{ M}^3 \times \frac{0.35 \text{ mg}}{M^3} = 7.93 \text{ mg Chlorpyrifos vapor in the bedroom atmosphere}$$
(at vapor saturation)

The 9.76 g of Chlorpyrifos in the gallon of paint greatly exceeds the 0.00793
g of total vapor in the bedroom's saturated atmosphere. Therefore, a vapor
saturation concentration was theoretically achievable, although highly
unlikely.

$$\frac{9.76 \text{ g}}{\text{gallon}} \times \frac{1000 \text{ mg}}{g} \times \frac{\text{gallon}}{3785 \text{ mL}} = \frac{2.58 \text{ mg}}{\text{mL}}$$

$$\frac{\dfrac{2.58 \text{ mg}}{M^3 - \text{hours}}}{8 \text{ hours}} = \frac{0.32 \text{ mg}}{M^3} = \text{eight-hour time-weighted average exposure to}$$

Chlorpyrifos mist

Since the paint density was 1.4 g/mL, the painter's eight-hr TWAE to total
mist = 1.4 g/mL × 1.0 mL × 1000 mg/g = 1400 mg total paint mist (includes
Chlorpyrifos):

$$\frac{\dfrac{1400 \text{ mg}}{M^3 - \text{hour}}}{8 \text{ hour}} = \frac{175 \text{ mg}}{M^3},$$ clearly a need for a full-face airline respirator during spraying.

Answers: 0.0246 ppm Chlorpyrifos vapor at saturation conditions. There was a sufficient amount of Chlorpyrifos in the paint to achieve vapor saturation in this bedroom. 0.044 mg Chlorpyrifos vapor/M^3 was the painter's calculated eight-hr TWAE. His/her total mist exposure, as an eight-hr TWAE, was 175 mg/M^3. In the author's opinion, insecticides, including organophosphates, should not be used in home interiors especially if infants and those with compromised health could be in enclosed rooms for long periods (e.g., sleeping eight or more hours, those with cholinesterase enzyme deficiencies, handicapped persons perpetually confined to bed, those with neurological impairments, pregnant women, etc.).

374. Which of the following situations is the most reasonable indication of a significant indoor air quality issue?

   a. Increase in relative humidity from 30 to 55%
   b. Decrease in relative humidity from 55 to 30%
   c. Hourly fluctuations in dry bulb temperatures from 70 to 55°F back to 70°F
   d. Decrease of air flow velocity at work stations from 200 to 100 fpm
   e. Absence of thermophilic actinomycetes in exhaust ducts and plenums
   f. Steady state air concentration of radon
   g. Increases of CO_2 from 300–400 ppm in the morning to > 1000 ppm by noon
   h. A stench of carbon monoxide gas in the air supply to employee work stations

Answer: g. Greater than 1000 ppm CO_2.

375. The vapor pressure of toluene 2,4-diisocyanate (TDI) is close to 0.01 mm of Hg at 77°F. The vapor pressure of methylene *bis*-phenyl diisocyanate (MDI) is about 0.001 mm of Hg at 104°F. Compare the relative volatility of these diisocyanates.

Answer: Since vapor pressure of a chemical approximately doubles with each 10°C (18°F) increase in temperature, it would be reduced by a factor of about two with each 18°F decrease in temperature. Therefore, as the temperature of liquid MDI is reduced by 18°F from 104°F to 86°F, MDI's vapor pressure drops from 0.001 mm Hg to about 0.0005 mm of Hg. If MDI's temperature is lowered another 18°F to 68°F, the vapor pressure of the liquid drops to nearly 0.00025 mm of Hg. So, by approximate interpolation, the vapor pressure of MDI at 77°F is nearly 0.00038 mm of Hg.

In other words, TDI is about $\dfrac{0.01 \text{ mm Hg}}{0.00038 \text{ mm Hg}}$ = 26.3 times more volatile than MDI. This is an oversimplification of relative volatility since other factors determine the evaporation rate of a chemical besides vapor pressure. Note that MDI is a solid below 99°F although it still exerts a vapor pressure in the solid phase. If the MDI covered a surface area approximately 26.3 times greater than the surface area covered by liquid TDI, one could say that the overall emission rate (flux) of the two diisocyanates would be nearly equal.

376. The American Industrial Hygiene Association Workplace Environmental Exposure Level (WEEL) for titanium tetrachloride is 500 micrograms per cubic meter of air. $TiCl_4$ rapidly hydrolyzes in moist air forming titanium dioxide fume and hydrogen chloride gas. Assuming stoichiometric conversion of the WEEL concentration to acid gas with an excess of water vapor in the atmosphere, what is the hydrogen chloride concentration? The molecular weight of titanium tetrachloride is 189.73.

$$TiCl_4 + 2 \ H_2O \rightarrow TiO_2 + 4 \ HCl$$

$$\dfrac{500 \times 10^{-6} \text{ g}}{189.73 \text{ g}} = 2.64 \times 10^{-6} \text{ mole of } TiCl_4$$
$$\text{mole}$$

Therefore, 4 (2.64 × 10^{-6} mol) = 10.56 × 10^{-6} mol of HCl is produced.

$$10.56 \times 10^{-6} \text{ mole HCl} \times \dfrac{36.5 \text{ g}}{\text{mole}} = 385.4 \times 10^{-6} \text{ g HCl} = 385.4 \text{ mcg HCl}$$

$$\text{ppm} = \dfrac{\dfrac{\text{mg}}{M^3} \times 24.45}{\text{molecular weight}} = \dfrac{\dfrac{0.3854 \text{ mg}}{M^3} \times 24.45}{36.5} = 0.26 \text{ ppm HCl}$$

Answer: 0.26 ppm HCl gas. This is below the ACGIH TLV (Ceiling) of 5 ppm. HCl gas is highly soluble in water and, therefore, tends to be more of an upper respiratory irritant than a deep lung toxicant. However, since the gas might adsorb on the tiny TiO_2 particles, there could be deep lung penetration since the fume is likely sub-micron in size.

377. Many finely-divided atmospheric dusts (e.g., flour, sugar, metals, starch, plastics) which settle and deposit on internal industrial plant structures can be a significant explosion hazard if a sufficient amount of dust becomes airborne and an ignition source is present. The amount of dust is generally considered to be significant if the settled amount exceeds:

   a. Any grossly visible amount removed by a piece of transparent adhesive tape
   b. 1/32 in.
   c. 1/16 in.

    d. 3/32 in.

    e. 1/8 in.

    f. 1/4 in.

    g. None of the above

Answer: e. The settled dust layer thickness should not exceed 1/8 in. If the settled dust exceeds this thickness, better process enclosure, exhaust ventilation, and plant housekeeping are required. "Cleaning" with compressed air wands must never be permitted.

378. Inert gases are often used to exclude oxygen from systems where combustible and flammable vapors could accumulate to explosive concentrations. Nitrogen, although not truly chemically inert, is the most common gas used for this purpose. For organic hydrocarbon vapors and gases (but not inorganic explosive gases and vapors such as hydrogen, carbon monoxide, and carbon disulfide), what is the minimum oxygen concentration for combustion?

    a. 21%

    b. 15%

    c. 10%

    d. 5%

    e. 3%

    f. none of the above

Answer: c. 10%, or slightly less than 0.5 atmosphere. In typical engineering practice, a safety factor is applied so that the actual oxygen concentration is 6% or less in carefully controlled atmospheres (such as drying ovens) and 2% or less in systems where poor mixing of diluent inert gas with combustible gas, for example, could occur. Besides nitrogen, other "inert" gases used for such purposes include argon, carbon dioxide, steam, and helium.

379. A test tube containing three mL of a culture of *L. pneumophillae* (720,000 organisms/mL) shatters in an ultracentrifuge. It is estimated that no more than 0.5 mL of this solution is released as a respirable aerosol into a 30 × 40 × 12 ft laboratory which has, for purposes of estimating the worst-case respirable dose, no ventilation. What is the average concentration of *L. pneumophillae* in the air?

$$\frac{720,000 \text{ organisms}}{\text{mL}} \times 0.5 \text{ mL} = 360,000 \text{ organisms}$$

$$30 \times 40 \times 12 \text{ ft} = 14,400 \text{ ft}^3 = 407,763 \text{ L}$$

$$\frac{360,000 \text{ organisms}}{407,763 \text{ liters}} = \frac{0.88 \text{ organism}}{\text{L}}$$

Answer: 88 *L. pneumophillae* bacterium per 100 L of air. It has been estimated that a single *L. pneumophillae* bacterium in 50 L of inhaled air is an infectious dose in a susceptible human host. A technician standing near the centrifuge might have been at greater risk since there would have been a burst microbial aerosol concentration considerably higher than 0.88 organism/L.

380. What is the density of dry air at 300°F and 760 mm Hg?

$$\text{air density} = \left[\frac{0.075\ \text{lb}}{\text{ft}^3}\right]\left[\frac{530°R}{460°R + 300°R}\right] = \frac{0.052\ \text{lb}}{\text{ft}^3}$$

Answer: 0.052 pound per cubic ft, or 69.3% that of dry air at NTP.

381. Three pounds of aluminum phosphide (AlP, molecular weight = 57.96 g/mol) are used to fumigate a 100,000 cubic ft grain storage building by the generation of phosphine gas (PH_3, molecular weight = 34.00 g/mol) from a quantitative reaction of the dry AlP pellets with atmospheric water vapor:

$$2\ \text{AlP} + 3\ H_2O \rightarrow 2\ PH_3\uparrow + Al_2O_3$$

What is the final concentration of phosphine gas in the building assuming the grain net void volume and headspace volume total 20% of the building volume?

three lb AlP = 1361 g AlP

1361 g AlP/57.96 g/mol = 23.48 mols of AlP. Therefore, since one mol of aluminum phosphide generates one mol of phosphine, 23.48 mols of PH_3 gas are released from the solid aluminum phosphide pellets.

23.48 mols × 34.00 g PH_3/mol = 798.3 g of PH_3

20% × 100,000 ft³ = 20,000 ft³ = 566.3 M³

$$\text{ppm} = \frac{\frac{\text{mg}}{M^3} \times 24.45}{\text{molecular weight}} = \frac{\frac{798,300\ \text{mg}}{566.3M^3} \times 24.45}{\frac{34.00\ \text{grams}}{\text{mole}}} = 1014\ \text{ppm}$$

Answer: 1014 ppm PH_3. TLV-TWA = 0.3 ppm. STEL = one ppm. The IDLH for PH_3 is 200 ppm. Application of three pounds of aluminum phosphide appears excessive, an "overkill," since the LC_{50} for rats (four hrs) is 11 ppm, and the LCL_os for rabbits = 2500 ppm for 20 min, for mice = 273 ppm for two hrs, for guinea pigs = 101 ppm for four hrs, and 50 ppm for two hrs for cats. The smallest amount of phosphine gas necessary to achieve sufficient vermin control should be chosen. An EPA-licensed phosphine fumigator could prescribe the safest effective dose for rodent and insect control. A legitimate question is if there is sufficient water vapor (relative humidity) in the air space to quantitatively convert AlP to phosphine. If not, one should ensure sufficient mechanical (water nebulizer) humidification of the atmosphere. Careful application of phosphine is warranted since human deaths have been associated with its use as a fumigant for insects in sealed railroad cars and other infested locations.

382. A four-in., full-port floating ball valve with self-relieving seats is in an outdoor line transporting dry, liquid chlorine at 500°F and 80 psia. This valve is inadvertently left closed when the line is drained and purged for maintenance. As a result, one liter of liquid chlorine is left trapped in the valve's body cavity. Piping on the downstream side had been removed for repairs. It is a hot and sunny day. Before long, the sun heats the valve up to 160°F increasing the cavity's pressure to 300 psig. The valve's seats are designed to relieve between 150 and 400 psia, and one seat does at 250 psia. The pressurized liquid chlorine immediately vaporizes forming a blast of gas. How much chlorine is released? In what air volume is produced a 1000 ppm cloud of chlorine gas at 760 mm Hg? Chlorine's liquid density at 160°F is 1237 kg/M³ = 1237 g/L. Molecular weight of chlorine = 70.906.

$$PV = nRT \quad 160°F = 344 \text{ K} \quad 760 \text{ mm H g} = \text{one atmosphere}$$

$$V = \frac{nRT}{P} = \frac{\left[\dfrac{1237 \text{ g/L}}{70,906 \text{ g/mole}}\right](0.0821 \text{ L - atm / K-mole})(344 \text{ K})}{1 \text{ atmosphere}} = 492.7 \text{ L}$$

As the pressurized, liquid chlorine rapidly blasts from the valve (perhaps injuring or killing somebody nearby), 491.7 L of gas are released (one liter of 100% Cl_2 gas remains in the valve body cavity for some time until it later slowly diffuses into the surrounding atmosphere).

$$1000 \text{ ppm} = 0.1\% \quad \frac{491.7 \text{ L}}{491.700 \text{ L}} \times 10^6 = 1000 \text{ ppm } Cl_2 \quad \frac{491.7}{0.001} = 491,700 \text{ L}$$

$$\sqrt[3]{491,700 \text{ L}} = \sqrt[3]{17,364 \text{ ft}^3} = 25.9 \text{ feet}$$

Answer: 491,700 L of chlorine gas are released. If confined to a volume 26 ft to a side, the cube would contain approximately an average of 1000 ppm of gas. The IDLH for chlorine is 30 ppm. Only a few inhalations of this gas concentration could be fatal to unprotected persons. If the valve rupture was closer to the ground (instead of a point-source release of gas), the spread of the gas cloud along the x and y axes would be greater than 26 ft; accordingly, the escape distance for unprotected workers would be longer than 26 ft. If released in a corner, for example, and under a low-lying roof or cover, the lateral escape distances become still longer. For example, if the gas is released in a corner under an eight-ft ceiling, the lateral distances in both of the horizontal dimensions become:

$$\sqrt{\frac{17,364 \text{ ft}^3}{8 \text{ ft}}} = \sqrt{2171 \text{ ft}^2} = 46.6 \text{ feet making safe evacuation less likely. The}$$

Cl_2 gas concentration in the corner near the valve would, of course, be greater than 1000 ppm. The blast effects might incapacitate a worker so greatly that he/she would be unable to evacuate the area.

383. Household chlorine bleach is a common, inexpensive, highly effective disinfectant often prescribed by industrial hygienists to sanitize and disinfect (but not sterilize) inanimate surfaces. Most commercial solutions as purchased are 5.25% sodium hypochlorite in water (NaOCl). It is thought that the ClO^- ions destroy bacteria by oxidizing life-sustaining compounds within them. How much stock bleach solution (@ 5.25% NaOCl) must be diluted with water to make one-gallon of a 3000 ppm disinfecting solution?

Have: 5.25% solution (w/v) = 5.25 g NaOCl/100 mL = 0.0525 g/mL

Want: 3000 ppm solution = 3000 mg NaOCl/L = 3 mg/mL = 0.003 g/mL

One gallon = 3785 mL

3785 mL/gallon × 0.003 g NaOCl/mL = 11.36 g NaOCl/gallon

$$\frac{11.36 \text{ g NaOCl}}{0.0525 \text{ g-mL NaOCl}} = 216.4 \text{ mL} = 7.3 \text{ ounces} = 0.92 \text{ cup}$$

Answer: Dilute almost one cup of household chlorine bleach to one gallon with water to make a 3000 ppm disinfecting solution.

384. A gas cylinder contains 1260 g of hydrogen fluoride at 17.5 atmospheres and 25°C. What mass of HF gas would be released if the cylinder was heated to 90°C, the valve was opened until the gas pressure dropped to one atmosphere, and the temperature was maintained at 90°C?

The HF jet gas volume is determined from the initial physical conditions:

$$n = (1260 \text{ grams}) \left[\frac{1 \text{ mole}}{20.01 \text{ grams}} \right] = 62.94 \text{ moles of HF}$$

$$V = \frac{nRT}{P} = \frac{(62.97 \text{ moles})(0.0821 \text{ L-atm/mole-K})(273 \text{ K} + 25 \text{ K})}{17.5 \text{ atmospheres}}$$

$$= 88.0 \text{ liters}$$

The final number of moles is:

$$n = \frac{PV}{RT} = \frac{(1.00 \text{ atm})(88.0 \text{ liters})}{(0.0821 \text{ L-atm/mole-K})(363 \text{ K})} = 2.95 \text{ moles of HF}$$

$$(2.95 \text{ moles HF}) \left[\frac{20.01 \text{ g}}{\text{mole}} \right] = 59.0 \text{ grams of HF}$$

$(1260 \text{ g HF initial}) - (59 \text{ g final}) = 1201 \text{ g of HF}$

Answer: 1201 g of HF gas escaped from the cylinder.

385. Calculate the composition of the vapor phase at 30°C which is in equilibrium with a solution of benzene (35 mol%) and toluene (65 mol%). The vapor pressures of these aromatic hydrocarbons at this temperature are 119 mm Hg and 37 mm Hg, respectively.

partial total vapor pressure from benzene = (0.35) (119 mm Hg) = 41.7 mm Hg

partial total vapor pressure from toluene = (0.65) (37 mm Hg) = 24.1 mm Hg

total vapor pressure = 41.7 mm Hg + 24.1 mm Hg = 65.8 mm Hg

The percent vapor composition is calculated by applying Dalton's law of partial pressures:

$$\text{concentration}_{\text{benzene}} = \frac{VP_{\text{benzene}}}{VP_{\text{total}}} = \frac{41.7 \text{ mm Hg}}{65.8 \text{ mm Hg}} = 0.634 = 63.4\% \text{ benzene}$$

$$\text{concentration}_{\text{toluene}} = \frac{VP_{\text{toluene}}}{VP_{\text{total}}} = \frac{24.1 \text{ mm Hg}}{65.8 \text{ mm Hg}} = 0.366$$

$$= 1.000 - 0.634 = 36.6\% \text{ toluene}$$

Answers: The vapor phase at saturation is 63.4% benzene and 36.6% toluene. Note how the more volatile component (benzene with its higher vapor pressure) is enriched from 35% in the liquid phase to 63.4% in the vapor phase.

386. The relative variability of the randomly distributed errors in a normal distribution is referred to as the coefficient of variation (CV). In air sampling, there are numerous opportunities for errors (analysis, sample collection, air flow rate calibration, etc.). Collectively, there can be a CV for the combined errors. For the following sampling and analytical procedures, rank their respective CVs: colorimetric detector tubes, rotameters on personal pumps, charcoal tubes, asbestos, respirable dust, gross dust.

Answers:		
Rotameter on personal pumps (sampling only)	0.05 CV
Gross dust (sampling/analytical)	0.05
Respirable dust, except coal mine dust (sampling/analytical)	0.09
Charcoal tubes (sampling/analytical)	0.10
Colorimetric detector tubes	0.14
Asbestos (sampling/counting)	0.24–0.38

387. An 8.3 L/min (nominal) critical orifice was calibrated in Philadelphia at 8.9 L/min at 72°F. This critical orifice was used to collect a general area air sample for silica dust at 45°F in a foundry also in Philadelphia. What was the actual air flow rate through the critical orifice? Assume identical barometric pressures for calibration and air sampling conditions.

$$Q_{actual} = Q_{indicated}\sqrt{\frac{T_{actual}}{T_{calibration}}} \quad Q_{actual} = \sqrt{\frac{460° + 45°}{460° + 72°}} = 8.67\,L/min$$

Answer: 8.67 L of air/min. If the barometric pressures were different between the calibration and sampling conditions, use this standard formula:

$$Q_{actual} = Q_{indicated}\sqrt{\frac{P_{calib} \times T_{actual}}{P_{actual} \times T_{calib}}} \quad \text{Always ensure that all units are the same and}$$

that temperatures are expressed in degrees absolute. Note that the square root function is always used with variable orifice fluid flow meters.

388. 700 cubic yards of dry, densely-packed cement dust are transferred approximately every hour during a three-min period from the calciner drying kiln into a storage silo. The procedure requires the operator to turn on a fan for the mechanical local exhaust system prior to dumping and allowing it to run as cement dust cascades into the silo. Displaced dusty air exits the silo through a 3 × 5 ft hole. An engineer selected a dust ventilation capture velocity of 300 ft/min for the hole face. Is this sufficient to prevent escape of dust into the surrounding atmosphere? The air void space in the packed cement dust (say, 10% of the total volume) does not enter into calculations since this air, along with the cement dust, is expelled from the silo.

700 yd³ × 27 ft³/yd³ = 18,900 ft³

18,900 ft³/3 min = 6300 ft³ of dusty air are displaced from the silo per min

Since the cement transfer process occurs in "burps" over the three min, allow a ventilation design contingency factor of five: 6300 ft³/min × 5 = 31,500 cfm

$$Q = AV \quad V = \frac{Q}{A} = \frac{31,500\,cfm}{3\,ft \times 5\,ft} = 2100\,ft/minute$$

Answer: A capture velocity of 300 fpm is too low. With this velocity, 6300 cfm of dusty air are being displaced while the fan attempts to exhaust 4500 cfm (15 ft² × 300 fpm). The silo emits, on average, 1800 ft³ of dusty air/min if the 300 fpm capture velocity is selected. Choose the fan, dust collector, and appropriately-sized ducts to provide a silo exit air capture velocity of about 2100 ft/min. Decreasing the exhaust opening size increases capture velocity, or, alternatively, a smaller fan and exhaust system could be used with the 2100 cfm capture velocity.

This approach can also be used when designing and selecting mechanical local exhaust ventilation systems for drum and tank filling operations. For example, if it requires two min to fill a 55-gallon drum with volatile solvents (27.5 gal/min = 3.7 ft³/min), a marginal ventilation system would be able to capture the 3.7 ft³ of air saturated with solvent vapors every minute. In typical, good industrial hygiene practice, a larger volume is normally chosen such as, in this case, a 100 cfm exhaust "elephant trunk" placed within four in. of the bung hole.

389. Organic vapor emissions from point sources (VOC, or volatile organic carbon) can be determined in a variety of ways depending on the source, the hydrocarbon, and other fugitive vapor release factors. What formula can be used to estimate yearly vapor losses ("breathing loss") from volatile hydrocarbon storage tanks?

Answer: A good calculation procedure is found in the Environmental Protection Agency Publication *AP-42*:

$$L_B = 2.26 \times 10^{-2} M_v \left[\frac{P}{P_A - P} \right]^{0.68} D^{1.73} H^{0.51} \Delta T^{0.5} F_p C K_c$$

where:

L_B = organic vapor breathing loss, lb/yr
M_V = molecular weight of the vapor, lb/lb-mol
P = true vapor pressure, psia
P_A = average atmospheric pressure, psia
D = tank diameter, ft
H = average vapor space height, ft
T = average ambient diurnal temperature (Δ equals the daily change), °F
F_P = tank's paint factor, dimensionless
  (e.g., shiny aluminum vs. flat black)
C = adjustment factor for small tanks, dimensionless
K_C = product factor, dimensionless

The EPA publication provides data on common hydrocarbons and various emission factors and coefficients. *EPA-405/4-88-004* and *EPA-450/2-90-001a* are other useful publications.

The working loss of a tank is the amount of volatile vapors displaced during loading and unloading the tank. Assuming that there is no vapor recovery system, this can be estimated by the equation:

$$L_W = 2.40 \times 10^{-5} M_v P V N K_N K_C$$

where:

L_W = working loss, lb/yr
V = tank capacity, gallons
N = number of capacity turnovers during the year, dimensionless
K_N = turnover factor from *AP-42*

390. Silane (SiH_4, silicon tetrahydride) has a TLV of five ppm. This pyrophoric gas readily reacts with water vapor in the atmosphere according to the equation:

$$SiH_4 + 2\ H_2O \rightarrow SiO_2 + 4\ H_2 .$$

Assume the atmosphere in a 20,000 gallon confined space contains 10,000 ppm SiH_4 vapor (1%). There is sufficient water vapor present to quantitatively hydrolyze the silane to silicon dioxide aerosol and hydrogen gas. Is an explosive atmosphere generated (LEL $H_2 \cong 4\%$ in air)? Molecular weight of silane is 32.12. Assume the heat of chemical reaction is sufficient to ignite hydrogen gas. Silane is a colorless, spontaneously flammable and pyrophoric gas. Most likely, silane vapor reacts with atmospheric water vapor to form silicic acid (hydrated silica, $SiO_2.n\text{-}H_2O$). The formation of crystalline silica in an aerosol cloud appears unlikely; amorphous silica appears to be the reaction product.

$$\frac{mg}{M^3} = \frac{ppm \times molecular\ weight}{24.45\ L/gram\text{-}mole} = \frac{10,000\ ppm \times 32.12\ g/mole}{24.45\ L/g\text{-}mole}$$

$$= 13,137\ mg/M^3$$

20,000 gallons = 75.71 M^3 (13,137 mg/M^3) $\times$ 75.71 M^3 = 994,602 mg SiH_4

$$\frac{994.6\ grams\ SiH_4}{\dfrac{32.12\ grams\ SiH_4}{mole}} = 30.97\ moles\ of\ SiH_4\ are\ in\ the\ tank's\ atmosphere$$

Since one mol of silane generates four mols of hydrogen gas, 30.97 mols will produce 123.88 mols of hydrogen. 123.88 mols $\times$ 2.016 g/mol = 247.7 g of hydrogen gas.

$$ppm = \frac{\dfrac{247,700\ mg\ H_2}{75.71\ M^3} \times 24.45}{2.016\ g/mole} = 39,679\ ppm\ H_2$$

Answer: 39,679 ppm hydrogen gas. This equals 3.97% or right at the lower explosive level of 4% hydrogen in air. While one debates if there is quantitative conversion of silane to hydrogen, since silane has a "built-in match", this tank and much nearby shortly will become history. Prevention of silicosis is not a key issue here.

391. What is the average vapor concentration after one cup of benzene evaporates into a building the size of a football field with a 14-ft ceiling? Assume that there is no dilution, only mixing, ventilation as the benzene evaporates. In

other words, what is the equilibrium vapor concentration within the building assuming no dilution?

$300 \times 160 \times 14 \text{ ft} = 672,000 \text{ ft}^3 = 19,030 \text{ M}^3$

one cup = 236.6 mL

$236.6 \text{ mL} \times 0.88 \text{ g/mL} = 208.208 \text{ g} = 208,208 \text{ mg}$

$$\text{ppm} = \frac{\dfrac{\text{mg}}{\text{M}^3} \times 24.45}{\text{molecular weight}} = \frac{\dfrac{208,208 \text{ mg}}{19,030 \text{ M}^3} \times 24.45}{78.11} = 3.4 \text{ ppm}$$

Answer: 3.4 ppm benzene vapor, or 6.6 times the TLV. Such calculations of hypothetical examples are helpful to explain low air contaminant concentrations to workers, supervisors, and to lay persons, such as jurors, the courts, your spouse and children, and to the neighbors. Using another familiar example — a standard sewing thimble of 2.2 mL volume: spill this amount of benzene inside a 1 1/2 car garage ($14 \times 20 \times 8$ ft), and the resultant vapor concentration becomes 9.5 ppm — almost 10 times the OSHA PEL (assuming, of course, the garage is unventilated).

392. What formula can be used to determine the amount of oxygen required for perfect combustion of any fuel?

Answer: The approximate theoretical volume of oxygen required to burn a fuel containing carbon, hydrogen, sulfur, and oxygen is:

$$\frac{\text{volume of oxygen}}{\text{pound of fuel}} = 359 \text{ ft}^3 \text{ or } 1710 \text{ ft}^3 \left[\frac{C}{12} + \frac{H_2}{4} + \frac{S}{32} - \frac{O_2}{32} \right]$$

The coefficient of 359 is the volume (in ft³) of one pound-mol of O_2 at 32°F and one atmosphere of pressure. Use 1710 as the coefficient to obtain the theoretical volume of air. This formula is used for any dry fuel (e.g., oil, coal, wood, gasoline, propane, kerosene, etc.). C, H_2, S, and O_2 are decimal weights of these elements in one pound of fuel (e.g., for a fuel with 81% carbon, C = 0.81).

393. Flue gases (containing SO_2) at 280°F are fed to a power plant stack through a three-ft diameter duct. The centerline Pitot tube measurement is 0.7 in. of water. The static pressure duct wall manometer reads –0.6 in. of water. The gas contains 0.45 mol-% SO_2. The barometric pressure is 725 mm Hg. What is the emission rate of SO_2 gas from this stack in pounds per hour? Assume that the Pitot traverse and static pressure gauge exceed 10 duct diameters downstream of any gas flow disturbances.

$$\Delta P = hp = \left[0.7 \text{ inch} - (-0.6 \text{ inch}) \right] \times \frac{1 \text{ ft}}{12 \text{ in}} \times \frac{62.4 \text{ lb}}{\text{ft}^3} = \frac{6.76 \text{ lt}}{\text{ft}^3}$$

$$P_{air} = \frac{1 \text{ lb-mole}}{359 \text{ ft}^3} \times \frac{29 \text{ lb}}{mole} \times \frac{492°R}{740°R} \times \frac{725 \text{ mm Hg}}{760 \text{ mm Hg}} = \frac{0.05123 \text{ lb}}{ft^3}$$

$$\frac{v_{max}^2}{2 g_c} = \frac{\Delta P}{r} = h_v \qquad v_{max} = \sqrt{2 g_c \times \frac{\Delta P}{r}} = \sqrt{\frac{(2)(32.2)(6.76)}{0.05123 \text{ lb}/ft^3}}$$

$$v_{max} = \frac{92.2 \text{ ft}}{second} \qquad \frac{v_{avg}}{v_{max}} = 0.81 \qquad v_{avg} = (0.81)(92.2 \text{ ft/second}) = 74.7 \text{ ft/second}$$

$$Q = A \, v_{avg} = \frac{\pi}{4}(3 \text{ ft})^2 \times \frac{74.7 \text{ ft}}{second} = \frac{528 \text{ ft}^3}{second}$$

$$SO_2 \text{ gas emission rate} = (528)(0.05123)(3600 \text{ seconds/hour})\left(\frac{1 \text{ lb-mole}}{29 \text{ lb}}\right)$$

$$\left(\frac{64 \text{ lb } SO_2}{\text{lb mole}}\right)(0.0045) = \frac{967 \text{ lb } SO_2}{hour}$$

Answer: 967 pounds of SO_2 gas per hr.

394. A gas mixture containing hydrogen chloride and dry air at 22 psia and 22°C is bubbled through a gas scrubber containing 180 mL of 0.01 N NaOH solution. The collection efficiency of the gas bubbler is 100%. The remaining NaOH is back-titrated requiring 15.63 mL of 0.1 N HCl. The dry air mixture has a wet test meter volume reading of 1.0 L at a pressure of 740 mm Hg and 25°C. What is the mass fraction and the mole fraction of HCl gas in the original gas mixture?

NaOH + HCl → NaCl + H$_2$O

Mols:
NaOH at start = (0.18)(0.01) = 0.0018 mol
NaOH at end = (0.01563)(0.1) = 0.001563 mol
HCl scrubbed = Δ NaOH$_{start}$ – NaOH$_{end}$ = 0.000237 mol

Assume no water vapor in the air when volume was measured in wet test meter:

$$\text{moles of air} = \frac{PV}{RT} = \frac{\left[\dfrac{740 \text{ mm Hg}}{760 \text{ mm Hg}}\right](1000 \text{m L})}{(82.06)(298)} = 0.0398 \text{ mole}$$

Assume water vapor in air:
$P_{H_2 0} = 0.4594 \text{ psia} = 23.76 \text{ mm Hg}$
$P_{air} = 740 \text{ mm Hg} - 23.76 \text{ mm Hg} = 716.24 \text{ mm Hg}$

$$\text{moles of air} = \frac{PV}{RT} = \frac{\left[\frac{716.24}{760}\right](1000\text{ mL})}{(82.06)(298)} = 0.0385 \text{ mole}$$

Water vapor in air:

$$Y_{HCl} = \frac{0.000237}{0.000237 + 0.0385} = 0.0061$$

$$Y_{HCl} = \frac{Y_{HCl}\left[\dfrac{M_{HCl}}{M_{air}}\right]}{1 + y_{HCl}\left[\dfrac{M_{HCl}}{M_{air}} - 1\right]} = \frac{(0.0061)\left[\dfrac{36.5}{29}\right]}{(1 + 0.0061)\left[\dfrac{36.5}{29} - 1\right]} = 0.0295$$

No water vapor in air:

$$y = \frac{0.000237}{0.000237 + 0.0398} = 0.0059 \quad y = \frac{(0.0059)\left[\dfrac{36.5}{29}\right]}{(1 + 0.0059)\left[\dfrac{36.5}{29} - 1\right]} = 0.0286$$

Answers: 0.0295 and 0.0286 mol fractions of HCl gas, respectively, in moist and in dry air (0.0295 and 0.0286 mol HCl, respectively/liter of moist or dry air). Converting, for example, 0.0295 mol of HCl/L into ppm:

$$0.0295 \text{ mol HCl} \times 36.46 \text{ g HCl/mol} = 1.076 \text{ g HCl} = 1{,}076{,}000 \text{ mcg HCl}$$

$$\frac{\dfrac{1{,}076{,}000 \text{ mcg HCl}}{L} \times 24.45 \text{ L/g-mole}}{36.46 \text{ g/g-mole}} = 721{,}563 \text{ ppm HCl} = 72.16\%$$

Such an obviously high concentration of HCl in this gas sample shows that it is not a workplace air sample. This is the concentration one might expect in a chemical process stream, say in the product effluent of a HCl-manufacturing plant: $Cl_2 + H_2 \rightarrow 2\text{ HCl}$. A one-L air sample from such a process stream is normally sufficient.

395. The concentration of iron oxide fume (as Fe_2O_3) in the breathing zone of a scarfer of new steel ingots is 4.7 mg/M³. The TLV for iron oxide fume is five mg/M³ (as Fe). What is the exposure of this scarfer as Fe?

The molecular weights of Fe_2O_3 and Fe = 159.69 and 55.847, respectively.

$$\frac{2\text{ Fe}}{Fe_2O_3} = \frac{2 \times 55.847}{159.69} = 0.6994 = 69.94\% \text{ Fe}$$

$$4.7 \text{ mg } Fe_2O_3/M^3 \times 0.6994 = 3.29 \text{ mg Fe}/M^3$$

Answer: 3.29 mg Fe_2O_3 per cubic meter.

396. The static pressure at a fan outlet is 0.5 in. of water. The static pressure at the fan's inlet is –6.5 in. of water. The duct velocity pressure is 0.8 in. of water. What is the fan static pressure?

$FSP = SP_{outlet} - SP_{inlet} - VP = +0.5$ in. $- (-6.5$ in.$) - 0.8$ in. $= 6.2$ in. of water

Answer: 6.2 in. of water.

397. Carbon dioxide gas measurements are used to estimate the percent of outdoor air being admitted to a building where the outdoor air is 400 ppm CO_2, the return air is 750 ppm CO_2, and the mixed air is 650 ppm CO_2. What is the percent outdoor air supplied under these conditions?

$$\% \text{ outdoor air} = \frac{ppm_{return\ air} - ppm_{mixed\ air}}{ppm_{return\ air} - ppm_{outdoor\ air}} \times 100 = \frac{750\ ppm - 650\ ppm}{750\ ppm - 400\ ppm} \times 100 =$$

$$\frac{100\ ppm}{350\ ppm} \times 100 = 28.6\%$$

Answer: Approximately 29% of the building's general ventilation air is supplied from outside. About 71% of the air in the building is being recirculated.

398. A 13,500 gallon liquid incinerator feed tank is filled and emptied daily with a PCB-contaminated solution of ethyl acetate (100 ppm PCBs). The tank is vented to the atmosphere and does not have a vapor recovery system. Does the incinerator or the tank release more hydrocarbons to the atmosphere? The vapor pressure of ethyl acetate at NTP is 73 mm Hg. The vapor pressure of PCBs, compared to the vapor pressure of ethyl acetate, is negligible. The design combustion efficiency of this incinerator is verified at 99.99%. The specific gravity of ethyl acetate is 0.90 g/mL. The molecular weight of ethyl acetate = 88.1 g/g-mol.

$$\text{partial pressure of ethyl acetate in tank} = \frac{73\ \text{mm Hg}}{760\ \text{mm Hg}}$$

$$= 0.096\ \text{atmosphere at NTP}$$

$$\text{vapor concentration in tank} = \frac{PM}{RT}$$

$$= \frac{(0.096\ \text{atm})(88.1\ \text{lb} / \text{lb-mole})}{0.7302\ \text{atm-ft}^3 / \text{lb-mole} - °R\ (460 + 75)}$$

$= 0.0217\ \text{lb/ft}^3$

The daily emission rate from the tank's open vent =

$$\frac{13,500\ \text{gallons}}{\text{day}} \times \frac{0.134\ \text{ft}^3}{\text{gallon}} \times \frac{0.0217\ \text{lb}}{\text{ft}^3} = \frac{39.3\ \text{lb}}{\text{day}}$$

The daily emission rate from this incinerator =

$$\frac{13,500 \text{ gallons}}{\text{day}} \times 0.90 \times \frac{8.34 \text{ lb}}{\text{gallon}} \times (1 - 0.9999) = \frac{10.1 \text{ lb}}{\text{day}}$$

Answer: The tank's vent emits almost four times the amount of vapors released from the incinerator's stack. Engineering controls for the tank include an inert gas blanket (nitrogen) to reduce explosion hazards. Ethyl acetate vapors pass through a regenerative carbon adsorber as the tank is filled. Do not vent these vapors into the incinerator. The present combined daily VOC emissions for this system is 39.3 lb plus 10.1 lb = 49.4 pounds. *Note:* If incinerator efficiency drops from 99.99% to 99.95%, the incinerator's contribution of VOC to the environment exceeds the fugitive VOC emissions from the tank by 50.7 lb/day – 39.3 lb/day = 11.4 lb/day.

$$\frac{13,500 \text{ gallons}}{\text{day}} \times 0.90 \times \frac{8.34 \text{ lb}}{\text{gallon}} \times (1 - 0.9995) = \frac{50.7 \text{ lb}}{\text{day}}$$

399. Leakage of highly and acutely toxic gases (e.g., arsine) from pressurized systems, especially into confined work spaces with poor ventilation, is problematic for the chemical safety engineer and workers in those spaces. Dish detergent solutions and commercial leak detection liquids can be used to determine the volumetric flow rate of gas leaks. The emission rate can be estimated with a small ruler scale and a stopwatch. The diameter of an enlarging gas bubble is measured as a function of time of growth. The assumption is made that the bubble is a sphere. Calculate the leak rate of 12% phosgene gas in nitrogen if a two cm diameter bubble is emitted from a gas line fitting in 7.3 sec.

$$V = \frac{\frac{1}{6} \times \pi \times d^3}{t} = \frac{\frac{1}{6} \times \pi \times (2 \text{ cm})^3}{7.3 \text{ seconds}} = \frac{0.574 \text{ cm}^3}{\text{second}} = \frac{0.0344 \text{ L}}{\text{minute}} = \frac{0.0122 \text{ ft}^3}{\text{minute}}$$

$$\frac{0.0122 \text{ ft}^3}{\text{minute}} \times 0.12 = \frac{0.001464 \text{ ft}^3 \text{ COCl}_2}{\text{minute}},$$

where:

V = gas leak rate, volume per time
d = diameter of gas bubble at time t, and
t = elapsed time to form the gas bubble with diameter, d

Answer: Approximately 0.001464 ft³ of phosgene gas is emitted per minute. If the leak is not repaired immediately or mechanical local exhaust ventilation is not readily available, then the volume of dilution air required to reduce phosgene gas to a safer concentration can be calculated. In general, it is poor industrial hygiene practice to rely upon dilution ventilation for highly toxic gases. In this example, the amount and rate of gas emitted appear to be relatively tiny; however, any leak of a highly toxic gas, such as phosgene, should be staunched immediately.

400. Helium, a gas considerably less dense than air (0.0103 lb/ft³; air = 0.075 lb/ft³), is used, among other purposes, to inflate party balloons. Most of us know that when one inhales helium gas, our voice becomes "squeaky" due to a lower density of the helium and air gas mixture passing over vocal cords in the larynx. Fatalities have resulted when persons have deliberately inhaled helium released from cylinders of pressurized gas. Deaths have been attributed to physical asphyxiation from the gas and from increased intrapulmonary pressure. Direct inhalation of helium from a commercial balloon-filling system can pose a greater hazard than inhaling helium from a party balloon. How can pressure of the inhaled helium gas dispensed from a commercial system kill one essentially instantly?

A difference of 30 mm Hg pressure between the surrounding lung pressure and the intrapulmonary pressure can be fatal if the exposure is prolonged. As the helium (or any inhaled gas) pressure differential increases to 80 to 100 mm Hg, death is rapid from rupture of alveoli. Prompt hemorrhage results in drowning asphyxiation fatalities.

Commercial helium balloon-filling systems typically deliver gas at five cubic feet per min (2.36 L/min). Lung volumes vary from person to person. Total lung capacity for an average adult man is 5.6 L, and, for an average adult woman, it is 4.4 L. For this evaluation, assume that a 10-year old child with a total lung volume of three L directly inhaled helium from a commercial gas-dispensing system.

lung bursting pressure = 80 to 100 mm Hg = 1.55 to 1.93 lb/in² (psi)

As an added measure of safety, use the lower limit of 1.55 psi for lung rupture. The questions become: 1. How much gas volume must be added to the total lung volume to cause a pressure increase of 1.55 psi?, and, 2. How fast could this occur?

Boyle's gas law applies:

$$\frac{P_1}{P_2} = \frac{V_2}{V_1 + v} \ ,$$

where P_1 and P_2 = original and final gas pressures, respectively; and V_1 and V_2 = the original and final gas volumes, respectively. v = the additional gas volume to be added to a three liter lung volume to cause a pressure increase of 1.55 psi. Atmospheric pressure is assumed to be 14.7 psi. Pressure at other altitudes can be substituted for this sea level pressure.

$$\frac{14.7\,\text{psi}}{14.7\,\text{psi}+1.55\,\text{psi}}=\frac{3.0\,\text{L}}{3.0\,\text{L}+v} \qquad (3.0\,\text{L})(16.25\,\text{psi})=(14.7\,\text{psi})(3.0\,\text{L}+v)$$

$v = 0.316$ L, the additional lung volume required

The minimum time required to produce this additional lung volume at the rupture pressure (1.55 psi) is calculated by dividing the additional lung volume of 0.316 L by the maximum helium gas flow rate, or:

$$\frac{0.316\,\text{L}}{\dfrac{2.36\,\text{L}}{\text{second}}} = 0.134 \text{ second}$$

Answer: The increased inhaled gas volume in a person with this lung capacity is only 0.316 L, and the pressure increase time is no less than 0.134 second. But, since lungs are compliant and are not a rigid structure, the time required to reach this increased pressure may be a tad longer. Regardless, these calculations show that inhalation of helium gas directly from portable balloon-filling tanks is extremely hazardous. And, since these portable tanks have initial pressures above 200 psi, there is ample pressure to cause fatal injuries in persons who place their mouth directly on the helium discharge gas valve.

401. A chemical reaction is predicted to release 560 cubic meters of process air which will contain 4390 ppm of H_2S. A combination of sodium hydroxide (NaOH) solution followed by a sodium hypochlorite (NaOCl) solution oxidation will be used for the removal of H_2S from the process gas stream. The H_2S neutralization and oxidation reactions are:

$$H_2S + 2\,NaOH \leftrightarrow Na_2S + 2\,H_2O$$

$$Na_2S + 4\,NaOCl \rightarrow Na_2SO_4 + 4\,NaCl$$

The H_2S gas will be scrubbed through a column containing 55 gallons of 1% NaOH solution in water. It is assumed that, if there is sufficient NaOH present, the H_2S gas absorption will be 100% efficient. The resultant sodium sulfide solution is then reacted with the sodium hypochlorite solution for final oxidation. Is there sufficient NaOH present?

$$\frac{mg}{M^3} = \frac{ppm \times \text{molecular weight}}{24.45} = \frac{4390\,ppm \times 34.08}{24.45} = \frac{6119\,mg\,H_2S}{M^3}$$

$$\frac{6119\,mg\,H_2S}{M^3} \times 590\,M^3 = 3{,}610{,}210\,mg\,H_2S = 3610 \text{ grams of } H_2S$$

$$\frac{3610\,g\,H_2S}{\dfrac{34.08\,g\,H_2S}{\text{mole}}} = 105.9 \text{ moles of } H_2S \text{ are to scrubbed}$$

Therefore, 105.9 mols $\times$ 2 = 211.8 mols of NaOH are needed for 100% reaction.

55 gallons = 208.2 L.   1% (w/v) solution = 10 g NaOH/L

208.2 L $\times$ 10 g NaOH/L = 2082 g NaOH

$$\frac{2082 \text{ g NaOH}}{\frac{40.00 \text{ g NaOH}}{\text{mole}}} = 52.05 \text{ moles of HaOH! } 211.8 \text{ moles of NaOH required}$$

Answer: There is insufficient NaOH present in the gas scrubber solution for the removal of H_2S from the process stream. Increase the concentration to at least 5% to ensure stoichiometric reaction. A 5% solution contains 260.25 mols of NaOH, or 48.45 mols in excess. Consideration should be given to scrubbing the gas process stream through two absorption columns connected in series with the first containing 55 gallons of a 4% solution of NaOH, and with the second containing 55 gallons of a 2% solution of NaOH. After completion of the reaction, combined scrubber solutions are oxidized with sodium hypochlorite. The use of a flare to oxidize the H_2S or injection through a high level stack to dilute the gas to safer ground-level concentrations is risky, possibly much more costly, and likely prohibited. Issues of inhalation toxicity and community odor would remain. In 1950 in Poza Rica, Mexico, a flare failed to oxidize H_2S to SO_2 gas at a sulfur recovery unit in a refinery, and, during the night, 22 persons in the community died with another 320 hospitalized. H_2S is a potent chemical asphyxiant with an IDLH of 100 ppm. CO and HCN, two other chemical asphyxiant gases, have IDLHs of 1200 ppm and 50 ppm, respectively.

402. The concentration of radon-222 in the basement of a newly-constructed house is 1.85 $\times$ 10^{-6} m/L. Assuming that the air in the basement remains static, and there is no further release of Rn-222, what is the concentration after 2.3 days? The half-life of Rn-222 is 3.8 days.

$$\text{amount of Rn-222 remaining} = \left(1.85 \times 10^{-6} \text{ mole/L}\right)\left[\frac{1}{2}\right]^{\frac{2.3 \text{ days}}{3.8 \text{ days}}} =$$

$$\left(1.85 \times 10^{-6} \text{ mole/L}\right)\left[\frac{1}{2}\right]^{0.605} = 1.22 \times 10^{-6} \text{ mole/L}$$

Answer: 1.22 $\times$ 10^{-6} mol per liter, or (1.22 $\times$ 10^{-6} mol/L) $\times$ (222 g Rn/mol) = 270.84 $\times$ 10^{-6} g/L = 271.84 mcg/L.

403. 10.9 L of ambient, community air at 740 mm Hg and 19.0°C were bubbled through lime water [an aqueous suspension of $Ca(OH)_2$]. All carbon dioxide gas in the air sample was precipitated as calcium carbonate. If the precipitate

weighed 0.058 g, what was the percent by volume of CO_2 in this air sample? The molecular weights of carbon dioxide and calcium carbonate are, respectively, 44.01 and 100.09. Disregard the very low solubility of calcium carbonate in water (0.00153 g/100 cc at 25°C).

$$CO_2 + Ca(OH)_2 \rightarrow CaCO_3{\downarrow} + H_2O \quad 10.9 \text{ L} = 0.0109 \text{ M}^3$$

$$\frac{\dfrac{0.058 \text{ g CaCO}_3}{100.09 \text{ g CaCO}_3}}{\text{mole}} = 0.00058 \text{ mole of CaCO}_3$$

Therefore, 0.00058 mol of CO_2 was in the 10.9 L air sample.

$$0.00058 \text{ mole of CO}_2 \times \left[\frac{44.01 \text{ g CO}_2}{\text{mole}} \right] = 0.0255 \text{ g of CO}_2 = 25.5 \text{ mg CO}_2$$

$$\text{ppm} = \frac{\text{mg}}{\text{M}^3} \times \frac{22.4}{\text{molecular weight}} \times \frac{\text{absol. temp.}}{273.15 \text{ K}} \times \frac{760 \text{ mm Hg}}{\text{pressure, mm Hg}} =$$

$$\frac{25.5 \text{ mg CO}_2}{0.0109 \text{ M}^3} \times \frac{22.4}{44.01} \times \frac{273.15°C + 19°C}{273.15} \times \frac{760 \text{ mm Hg}}{740 \text{ mm Hg}} = 1308 \text{ ppm CO}_2$$

Answer: 1308 ppm CO_2 = 0.1308%. This is obviously polluted air because the typical ambient concentration of CO_2 in relatively clean air is about 350 ppm. In such an atmosphere (most likely urban), there would most likely be several other air contaminants, e.g., CO, soot, HCHO, NO_x, SO_2, etc.

404. Inert gases, such as nitrogen, are used to exclude oxygen from systems where an explosion hazard exists and/or the product or materials are subject to oxidation. Ventilation formulae are presented in this book to enable the engineer to calculate how much nitrogen is needed to dilute an initial 21% ambient oxygen concentration down to some predetermined level, say <1% O_2. Devise some simple guidelines to enable a chemical process attendant to do this without resorting to complex 1st-order ventilation air contaminant decay equations.

Answer: The concentration of oxygen in the ambient air is 21%, so after seven inert gas volume exchanges in a system, the oxygen concentration decays to less than 0.1%. Thus, total nitrogen required for sweep purging would be seven times the headspace volume. For example, a 10,000 gallon tank containing 2000 gallons of liquid would require 7 × 8000 gallons = 56,000 gallons of nitrogen with good mixing of the sweep gas with the oxygen-containing air. Since fans are rated in cubic feet of air/min, convert gallons to cubic feet by multiplying by 0.134. 56,000 gallons × 0.134 = 7504 ft^3. Divide the rated fan capacity (in cfm) into total headspace volume to obtain fan operating time, e.g., with a 500 cfm fan blowing nitrogen into the space: 7504 ft^3/500 cfm = 15 min. Apply a safety factor if there is poor mixing of the nitrogen with the air. Verify the oxygen concentration by obtaining representative gas samples. In closed systems, a mass flow meter or a critical flow orifice can be used for measuring the nitrogen delivery.

People have died from asphyxiation when they unknowingly entered confined spaces which contained inert atmospheres. Such spaces must be clearly posted to warn of the inhalation hazard. All elements of confined space entry programs must be in place to conserve the health and safety of entrants (see 29 *CFR* 1910.146). Padlocking all inert atmosphere spaces to impede unauthorized entry is a tremendous idea; hazard warning signs are not enough.

405. An ambient air sample is collected in a partially-evacuated 3.2 L glass flask for gas analysis. The barometric pressure at the time of sampling is 728 mm Hg, and the air temperature is 25°C. The relative humidity is 45%. The pressure remaining in the flask after evacuation (but before air sampling) is 480 mm Hg. What air volume is sampled? The vapor pressure of water at 25°C (77°F) is 23.76 mm Hg.

The partial pressure of the air sample due to water vapor = 0.45 × 23.76 mm Hg = 10.7 mm Hg.

$$V = V_a \times \frac{P_{bar} - P_{H_2O} - P_{partial}}{760 \text{ mm Hg}} \times \frac{273°C}{273°C + 25°C} =$$

$$3.2 \text{ L} \times \frac{728 \text{ mm Hg} - (10.7 \text{ mm Hg} + 480 \text{ mm Hg})}{760 \text{ mm Hg}} \times \frac{273}{273 + 25} =$$

$$3.2 \text{ L} \times \frac{237.3}{760} \times \frac{273}{298} = 0.915 \text{ L, where:}$$

P_{bar} = barometric pressure

P_{H_2O} = partial pressure of water vapor

$P_{partial}$ = pressure remaining in the flask after evacuation

Answer: 0.915 L. Please note that these calculations correct gas sampling conditions to STP (0°C and 760 mm Hg).

406. Air was sampled for sodium nitrate dust and nitric acid gas at an average flow rate of 1.73 L/min through a membrane filter followed by an impinger which contained 9.7 mL of a dilute alkali. Assuming 100% collection efficiency and an air sampling duration of 29.5 min, how much dust and gas were present if the filter contained 161 mcg of NO_3^-, and the impinger contained 2.7 mcg of NO_3^-, per mL? The respective molecular weights of $NaNO_3$ and HNO_3 are 84.99 and 63.01. The molecular weight of nitrate anion, NO_3, is 62.01. Assume that this air sample was obtained at 25°C and 760 mm Hg.

It is reasonable to assume HNO_3 gas passed through the filter and was collected by the impinger and that essentially no sodium nitrate dust passed the filter.

1.73 L/min $\times$ 29.5 min = 51.035 L

$$161 \text{ mcg } HO_3^- \times \frac{84.99}{62.01} = 220.7 \text{ mcg } NaNO_3$$

$$\frac{220.7 \text{ mcg } NaNO_3}{51.035 \text{ L}} = 4.32 \text{ mg } NaNO_3/M^3$$

$$\left[9.7 \text{ mL} \times \frac{2.7 \text{ mcg } NO_3^-}{mL} \right] \times \frac{63.01}{62.01} = 26.6 \text{ mcg } HNO_3$$

$$\text{ppm} = \frac{\dfrac{26.6 \text{ mcg } NO_3}{51.035 \text{ L}} \times 63.01}{24.45 \text{ L/g-mole}} = 1.34 \text{ ppm } HNO_3$$

Answers: 4.33 mg $NaNO_3$ dust/M^3 and 1.34 ppm HNO_3 gas — a dusty, irritating atmosphere. The ACGIH TLVs for HNO_3 gas, total inhalable particulates (NOC), and respirable particulates are 2 ppm, 10 mg/M^3, and 3 mg/M^3, respectively. If we assume that one-third of the total airborne dust is respirable, what was the additive mixture exposure?

$$\frac{1.44 \text{ mg}/M^3}{3 \text{ mg}/M^3} + \frac{2.89 \text{ mg}/M^3}{10 \text{ mg}/M^3} + \frac{1.34 \text{ ppm}}{2 \text{ ppm}} = 1.44$$

clearly an overexposure if this air sample represented a worker's breathing zone and the exposure was prolonged. A short-term air sample, such as this, is perfectly acceptable to demonstrate the magnitude of exposures and if engineering and work practice controls are clearly required.

407. A worker exposed to a steady-state concentration of an air contaminant is required to wear an air-purifying respirator which has a protection factor of 100. What is the effective protection factor (P_{effect}) if he removes the respirator for only a 30-second period out of his eight-hr work day?

480 min $\times$ (60 sec/min) = 28,800 sec

$$\frac{30 \text{ seconds}}{28,800 \text{ seconds}} \times 100 = 0.104\% \text{ of the exposure time}$$

$1 - 0.104 = 99.986 = \%$ of exposure time when protection is provided

$$P_{effect} = \frac{100}{\left[\dfrac{99.986}{100} \right] + 0.104} = 90.59$$

$100 - 90.59 = 9.41$

Answer: Removing the respirator for only 30 sec during an eight-hr work day reduces the protection factor by 9.4%.

408. In the preceding example, what is P_{effect} if his respirator is worn 90% of the time?
     $480 \text{ min} \times 0.1 = 48 \text{ min}$

$$P_{effect} = \frac{100}{\left[\dfrac{90}{100}\right]+10} = 9.2 \qquad 100 - 9.2 = 90.8$$

Answer: If the respirator is removed for any combinations of time which total 48 min during an eight-hr work day, respirators with protection factors of 1000 and 100 will provide nearly a same level of respiratory protection. In this example, the respirator's protection factor is reduced by almost 91%.

409. A worker's exposure to an air contaminant released from a point emission source can be estimated using the dispersion calculation method. If an industrial process, for example, emits 33 milligrams of hydrogen fluoride gas per second from a point source (such as a leaking valve in a pressurized line) in a 10 mph horizontal wind flow, what is a worker's most likely exposure to this gas if he is 50 ft away from the point emission source (that is, downwind of the source)? What would it approximately be if he was 10 ft away?

$$C = \frac{Q}{k\,u\,x^n}, \text{ where:}$$

C = concentration of air contaminant expected in the breathing zone, mg/M^3

k = a constant, 0.136

u = wind speed, meters/second
    (with 0.5 m/sec being a minimum velocity)

x = distance between the worker and the source, meters

n = a constant, 1.84

Q = emission rate, milligrams/second

$10 \text{ mph} = 4.47 \text{ meters/second} \quad 50 \text{ ft} = 15.24 \text{ meters}$

$$C = \frac{33 \text{ mg HF/sec}}{(0.136)(4.47 \text{ m/sec})(15.24 \text{ meters})^{1.84}} = 0.36 \text{ mg/M}^3$$

$10 \text{ ft} = 3.048 \text{ meters}$

$$C = \frac{33 \text{ mg HF/sec}}{(0.136)(4.47 \text{ m/sec})(3.048 \text{ meters})^{1.84}} = 6.98 \text{ mg HF/M}^3$$

Answer: 0.36 mg HF/M³ at 50 ft away. The TLV (ceiling concentration) for HF is 2.3 mg/M³. At 10 ft, the predicted exposure concentration is seven mg HF/M³, or three × TLV (C). This illustrates a fundamental and primary industrial hygiene control method: reduce exposure by increasing the distance between the worker and environmental stressors (air contaminant, heat, noise, radiation, etc.).

410. Dry air at sea level has a total pressure of 760 mm Hg. Oxygen comprises 20.95% (159.22 mm Hg), and 79.05% (600.78 mm Hg) is comprised from nitrogen, argon, other inert gases, and carbon dioxide. If this air is humidified to 70% at 77°F (25°C), what is the change in the partial pressure of the gases of composition? Vapor pressure of water at 77°F is 23.76 mm Hg.

   0.7 × 23.76 mm Hg = 16.63 mm Hg (the partial pressure due to water vapor)

   for oxygen: 20.95% × (760 – 16.63) mm Hg = 155.73 mm Hg

   for nitrogen, argon, etc.: 79.05% × (760 – 16.63) mm Hg = 587.63 mm Hg

   Answers: 155.7 mm Hg for oxygen and 587.6 mm Hg for remaining gases.

411. The maximum allowable concentration of calcium carbonate dust (limestone) in a worker's breathing zone is selected to be 10% of the TLV = 1 mg/M³ (TLV = 10 mg/M³). The total general dilution ventilation through his workspace is 10,000 cfm with 20% of the air (2000 cfm) recirculated. The ventilation mixing factor, K, is estimated at 3. What would be the maximum allowable concentration of calcium carbonate dust in the discharge air from the dust collector before it is mixed with the room air to meet the 1 mg/M³ maximum exposure limit?

   $$C_R = \frac{1}{2}\left(TLV - C_o\right) \times \frac{Q_T}{Q_R} \times \frac{1}{K}, \text{ where:}$$

   C_R = concentration of contaminant in exit air from the collector before mixing

   Q_T = total ventilation through the affected workspace (ft³/min)

   Q_R = recirculated air flow (ft³/min)

   K = ventilation mixing factor, usually varying between three and 10 with one = excellent conditions, three = good mixing conditions, and 10 = extremely poor mixing conditions.

   TLV = threshold limit value of the air contaminant (only relatively nontoxic airborne contaminants should be considered for recirculation — and then with exquisite engineering controls)

   C_o = concentration of contaminant in worker's breathing zone with local exhaust discharged outside

$$C_R = \frac{1}{2} \times \left(10 \text{ mg/M}^3 - 1 \text{ mg/M}^3\right) \times \frac{10,000 \text{ cfm}}{2000 \text{ cfm}} \times \frac{1}{3} = 7.5 \text{ mg/M}^3$$

Answer: The maximum concentration in the discharge air of the collector can be no more than 7.5 mg/M³ (before mixing with dilution air) to ensure that breathing zone concentrations do not exceed one mg/M³.

412. A coal-burning power plant emits 0.2% SO_2 by volume at a temperature of 260°F from its stack. The total volume of gases emitted is 5×10^5 cubic feet per min. What is the SO_2 emission rate in units of mass/time (seconds)? The atmospheric pressure is 980 millibars.

molecular weight of $SO_2 = 64$ 260°F = 399.8 kelvin

500,000 cfm = 236 M³/second

$$\frac{\text{mass}}{\text{time}} = \frac{V}{t} \times \frac{PM}{RT} = \frac{\left(236 \text{ M}^3 / \text{second}\right)(0.002)(64)(980)}{(0.0832)(399.8 \text{ K})}$$

$$= 890 \text{ grams } SO_2/\text{second}$$

Answer: 890 g of SO_2 are emitted per second.

413. A boiler maker removing fly ash from an oil-fired furnace is exposed to an average total dust concentration of three mg/M³ over his eight-hr work shift. He does not use a respirator. Without an analysis of constituents of this fine dust, is this a potential significant inhalation hazard? In other words, should one not implement industrial hygiene controls because some information is lacking? After all, if this is simply a nuisance dust (TLVs: inhalable particulates = 10 mg/M³; respirable particulates = 3 mg/M³ if NOC), an over-exposure does not appear indicated.

<u>Typical Fuel Oil Ash Metal Analysis</u>

Iron	22.90 weight %
Aluminum	21.90
Vanadium	19.60 ★
Silicon	16.42
Nickel	11.86 ★
Magnesium	1.78
Chromium	1.37 ★
Calcium	1.14
Sodium	1.00
Cobalt	0.91 ★
Titanium	0.55
Molybdenum	0.23
Lead	0.17 ★

Copper	0.05
Silver	0.03
Miscellaneous	0.09
Total	100%

Source: *Air Pollution Engineering Manual*, U.S. Department of Health, Education, and Welfare (1967)

Applying the above typical percentages to just the ✱ highlighted metals:

3 mg/M³ × 0.196 (vanadium) = 0.59 mg V/M³   TLV = 0.05 mg/M³
                                               (as respirable
                                               V_2O_5 dust or fumes)

3 mg/M³ × 0.1186 (nickel) = 0.36 mg Ni/M³   TLV = 0.05 mg/M³
                                               (soluble Ni) and
                                               0.1 mg/M³ for
                                               (insoluble Ni)

3 mg/M³ × 0.0137 (chromium) = 0.04 mg   TLV = various:
  total chromium/M³   0.01 to 0.5 mg/M³
3 mg/M³ × 0.0091 (cobalt) = 0.03 mg Co/M³   TLV = 0.02 mg/M³
3 mg/M³ × 0.0017 (lead) = 0.005 mg Pb/M³   PEL = 0.05 mg/M³

Answer: Excessive exposures to several metals appear likely. Control methods are indicated. There might be crystalline silica (SiO_2) in the total silicon analysis.

414. Inert gas blanketing techniques are often used to help prevent explosions in closed systems. Nitrogen or steam are often used for this purpose. Vacuum purging with nitrogen gas is done by evacuating gases and vapors from the system and then increasing the system pressure back to atmospheric pressure with nitrogen. This is done repeatedly until the desired minimum oxygen concentration is achieved. What would the oxygen concentration be after three purges in a system where the attainable vacuum pressure is 0.6 atmosphere?

$O_n = 21\ P^n$, where:
O_n = oxygen concentration after n purges, and
P   = vacuum pressure, bar absolute.

0.6 atmosphere = 0.608 bar

$O_n = 21\ (0.608)^3 = 4.72\%$

To determine the number of purges to achieve a desired oxygen concentration:

$$n = \frac{\log\left[\dfrac{O_n}{21}\right]}{\log P} = (\text{from the preceding example}),\ \frac{\log\left[\dfrac{4.72\%}{21\%}\right]}{\log 0.608\ \text{bar}} = \frac{-0.648}{-9.216} = 3$$

Answer: 4.72% oxygen. This assumes that the inert gas does not contain any oxygen, and that the initial concentration of oxygen is 21%. Oxygen concentration should be regularly verified by a reliable gas analysis. These are oxygen deficient atmospheres without warning properties. Such systems must be posted, secured, and treated per OSHA's confined space entry procedure. If steam is used, special care must be exercised because many do not regard water vapor as an inert gas. The author investigated an asphyxiation fatality from steam because nitrogen was unavailable to create an inert head space blanket. The tank entrant, presumably, (and the site safety engineer) assumed that steam contained sufficient oxygen to support life. The tank atmosphere, tested after the worker's body was extracted, contained 2% oxygen. Refer to problem 417.

415. Secondary dust explosions can occur in buildings when an initial explosion from any cause entrains settled dust into the atmosphere, and the dust cloud is ignited. Assume that a small building 20 meters long, 10 meters wide, and five meters high has two millimeters of settled explosive dust dispersed evenly upon the floor. The bulk density of this dust is 450 kg/M³. If a small explosion in this building entrains 50% of the settled dust uniformly into the building's air, and there is a source of ignition, will there be a secondary explosion after the first blast wave?

50% of two mm bulk dust layer = one mm of settled dust suspended in the air

1 mm = 0.001 meter

floor area = 20 meters $\times$ 10 meters = 200 meters²

volume of dust suspended in air = 200 meters² $\times$ 0.001 meter = 0.2 M³

mass of dust suspended in air 0.2 M³ $\times$ 450 kg/M³ = 90 kg = 90,000 g

building volume = 20 m $\times$ 10 m $\times$ 5 m = 1000 M³

concentration of explosive dust in atmosphere $= \dfrac{90,000 \text{ g}}{1000 \text{ M}^3} = \dfrac{90 \text{ grams}}{\text{M}^3}$

Answer: 90 g of dust per cubic meter of air. Ranges of LELs for explosive dusts in air are approximately 10 to 2000 g per cubic meter. So, depending upon this particular dust, the blast wave from the initial explosion could entrain enough dust to result in a normally more substantial secondary dust explosion. For example, some minimum dust concentrations for explosion in air (g/M³) are: cornstarch (40), sugar (35), wheat flour (50), liver protein (45), aluminum stearate (15), coal (50), soap (45), rubber (25), aluminum (40), iron (250), zinc (480), and magnesium (10). Particle size of the dust (i.e., total surface area) is an important variable. Refer to problem 377.

416. Calculate the vapor/hazard ratio numbers for toluene and ethyl acetate. The vapor pressure and threshold limit value for toluene are, respectively, 21 mm Hg and 50 ppm, and for ethyl acetate, respectively, are 73 mm Hg and 400 ppm.

First, calculate the equilibrium saturation concentrations for each organic vapor:

for toluene: $\dfrac{21\,\text{mm Hg}}{760\,\text{mm Hg}} \times 10^6 = 27,632$ ppm

for ethyl acetate: $\dfrac{73\,\text{mm Hg}}{760\,\text{mm Hg}} \times 10^6 = 96,053$ ppm

Next, divide the equilibrium saturation concentrations for each organic vapor by its threshold limit value:

for toluene: $\dfrac{27,632\,\text{ppm}}{50\,\text{ppm}} = 553\,(\text{dimensionless})$

for ethyl acetate: $\dfrac{96,053\,\text{ppm}}{400\,\text{ppm}} = 240\,(\text{dimensionless})$

Answer: The vapor/hazard ratio for ethyl acetate is $553/240 = 2.3$ times below that of toluene indicating that, although it is more volatile than toluene and reaches higher concentrations more quickly, its lower inhalation toxicity makes it perhaps a better choice in selecting a solvent if fire issues are effectively controlled. Toluene, unlike ethyl acetate, is skin-absorbed, and it is a greater systemic toxicant. While the vapor/hazard ratio is a reasonably good approach to consider when selecting solvents, other industrial hygiene and chemical safety engineering factors must obviously be considered, e.g., flash point, ignition temperature, systemic *vs.* local toxic effects, carcinogenicity, teratogenicity, solvent mixtures, consequences and degrees of exposures, etc. This approach to risk management demonstrates that TLVs and PELs, alone, must not be used solely in arriving at a solvent selection.

417. Instead of vacuum purging to reduce the concentration of oxygen in a flammable atmosphere vessel (See problem 414), pressure purging can be performed. An inert gas, such as nitrogen, is used to pressurize the vessel. The pressure is then relieved in a safe area, and the process is repeated until a desirable minimum oxygen concentration is achieved. What is oxygen concentration in head space of a vessel after three pressure purges if the oxygen concentration of the purge gas is 3%, and the purge pressure is 2.3 bar?

$$O_n = O_p + \left(O_i - O_p\right)\left[\frac{1}{P^n}\right],\ \text{where:}$$

O_n = oxygen concentration after n purges,

O_i = initial oxygen concentration, normally taken as 21%,
O_p = oxygen concentration of the purge gas,
P = purge pressure, bar absolute, and
n = number of inert gas pressurizations.

$$O_n = 3\% \, (21\% - 3\%) \left[\frac{1}{2.3^3} \right] = 4.44\%$$

Answer: 4.44% oxygen. The number of pressure purges necessary to achieve a pre-selected minimum oxygen concentration can be calculated by rearranging the above equation:

$$n = \frac{\log\left[\dfrac{O_i - O_p}{O_n - O_p} \right]}{\log P} = (\text{using the example}), \frac{\log\left[\dfrac{21\% - 3\%}{4.44\% - 3\%} \right]}{\log 2.3}$$

$$= \frac{\log\left[\dfrac{18}{1.44} \right]}{\log 2.3} = 3.03$$

Pressure purging uses more inert gas than vacuum purging to achieve the same reduced oxygen concentration. An advantage of pressure purging over vacuum purging are the briefer cycle times; that is, it generally takes longer to develop vacuum than it does to pressurize a vessel. Cost and performance determine the best method which, in some cases, can be a combination of the two procedures. The vessel must be rated for the maximum pressure and/or vacuum before either method may be employed, and allow a safety design factor before proceeding.

The volume of inert gas required to reduce the oxygen level from O_1 to O_2 is:

$$Q\,t = V \ln\left[\frac{O_1 - O_o}{O_2 - O_o} \right], \text{ where:}$$

Q = volumetric flow rate,
t = time
V = vessel volume
O_o = inlet oxygen concentration
O_1 = initial oxygen concentration in vessel
O_2 = final oxygen concentration in vessel

418. Flow-through ventilation, or dilution purging, can be performed to create an inert (or reduced oxygen) atmosphere in a space with explosive vapors, gases, or dusts. The inert gas, usually nitrogen, flows longitudinally through the vessel and sweeps oxygen from the system. Choice of such a procedure requires careful placement of the inert gas entry point(s) and flow gas exit point(s).

What is the concentration of oxygen in the exit gas after 10 min if the initial concentration of oxygen in a tank was 21%, the inert gas flow rate is two cubic meters per second, the volume of the tank head space is 700 M³, and the oxygen concentration of the purge gas is 2%?

$$O_f = O_p + \left[\left(O_i - O_p \right) e^{\left[\frac{-Qt}{V} \right]} \right], \text{ where:}$$

O_f = oxygen concentration after time, t,
O_p = oxygen concentration of the inert purge gas,
O_i = initial oxygen concentration (usually taken as 21% O_2, or ambient),
Q = purge gas flow rate, and
V = head space volume of vessel.
10 min = 600 sec

$$Q_f = 2\% + \left[(21\% - 2\%) e^{\left[\frac{-2 \text{ M}^3/\text{second} \times 600 \text{ seconds}}{700 \text{ M}^3} \right]} \right] = 5.42\% \, O_2$$

Answer: 5.42% oxygen. This equation can be re-arranged to determine purge time to achieve the desired minimum oxygen concentration, in this case: 5.42%:

$$t = -\frac{V}{Q} \ln \left[\frac{O_f - O_p}{O_i - O_p} \right] = -\frac{700 \text{ M}^3}{2 \text{ M}^3/\text{second}} \ln \left[\frac{5.42\% - 2\%}{21\% - 2\%} \right] =$$

$$-(350 \text{ seconds}) \ln \frac{3.24}{18} = 600 \text{ seconds} = 10 \text{ minutes.}$$

419. What is the final relative humidity if air at 95°F dry bulb and 40% relative humidity is cooled to 70°F dry bulb without the addition or removal of moisture?

This can be determined from a psychrometric chart or calculated as follows:

Since a fixed mass of air is cooled, its volume will become smaller according to Charles' law: the volume of a mass of gas is directly proportional to its absolute temperature when the pressure is held constant, or, using 1000 ft³ as the original volume:

$$\frac{V_i}{T_i} = \frac{V_f}{T_f}, \text{ rearranging: } V_f = \frac{V_i T_f}{T_i} = \frac{\left(1000 \text{ ft}^3 \right) (294.26 \text{ K})}{308.15 \text{ K}} = 954.92 \text{ ft}^3$$

At 95°F, the vapor pressure of water is 42.18 mm Hg, and at 70°F, the vapor pressure of water is approximately 18.68 mm Hg.

A. ppm H_2O vapor at 95°F $= 0.4 \times \dfrac{42.18 \text{ mm Hg}}{760 \text{ mm Hg}} \times 10^6 = 22{,}200$ ppm

(100% relative humidity would be 55,500 ppm = 5.55% water vapor.)

The gram molar gas volumes at 760 mm Hg and 70°F and 95°F are, respectively, 24.13 L and 25.27 L.

$$\frac{\text{mg } H_2O}{M^3} = \frac{\text{ppm} \times \text{molecular weight}}{25.27} = \frac{22{,}200 \times 18}{25.27} = \frac{15{,}813 \text{ mg}}{M^3}$$

15,813 mg/M^3 = 447.776 mg/ft^3 at 95°F = 447,776 mg of water vapor in 1000 ft^3 at 70°F: 447,776 mg/954.92 ft^3 = 469 mg/ft^3 = 16,563 mg/M^3

$$\text{ppm} = \frac{\dfrac{16{,}563 \text{ mg}}{M^3} \times 24.13}{18} = 22{,}204 \text{ ppm}$$

Note how ppm are constant between the two conditions of temperature. Equation A (above) can be rearranged to solve for the relative humidity at the new condition:

% relative humidity =

$$\frac{\text{ppm } H_2O \text{ vapor } \left(\text{at elevated DB temperature}\right)}{\dfrac{\text{vapor pressure, mm Hg } \left(\text{at reduced DB temperature}\right)}{760 \text{ mm Hg}} \times 10^4} =$$

$$\frac{22{,}200 \text{ ppm}}{\dfrac{18.68 \text{ mm Hg}}{760 \text{ mm Hg}} \times 10^4} = 90.3\%$$

Answer: 90.3% relative humidity at 70°F — in other words, cooler, but moister air. Significant cooling (air conditioning) can be achieved if much of the moisture is removed from this air before it is supplied to occupied areas. If using psychrometric charts, note that the dew point temperature does not change between these two conditions.

420. A rectangular ventilation duct is 14 × 20 in.. What is the equivalent diameter size for a round duct?

$$D_{equiv} = (1.3) \left[\frac{(A \times B)^{0.625}}{(A+B)^{0.25}} \right], \text{ where:}$$

D_{equiv} = equivalent diameter of the round duct size for a rectangular duct, inches

A   = one side of the rectangular duct, inches

B   = adjacent size of the rectangular duct, inches

$$D_{equiv} = (1.3)\left[\frac{(14\times20)^{0.625}}{(14+20)^{0.25}}\right] = (1.3)\left[\frac{(280)^{0.625}}{(34)^{0.25}}\right] = 18.2 \text{ inches}$$

Answer: 18.2 in. Actually, select an 18-in. diameter duct since this size is essentially an "off-the-shelf" item. Selection of a smaller diameter increases duct transport velocity. In general, round ducts are stronger than un-braced rectangular ducts.

421. In a six-ft diameter duct, a 10-point Pitot tube traverse in each of two directions gave the following readings (in inches of water) and corresponding duct velocities calculated as shown below:

0.70 in. (4180 fpm)	0.62 in.	(3930 fpm)	gas temperature = 300°F
0.79 (4440)	0.65	(4030)	
0.83 (4550)	0.67	(4080)	altitude = 100 ft
0.89 (4710)	0.75	(4330)	
0.91 (4760)	0.90	(4740)	dew point of gas = 140°F
0.90 (4740)	0.89	(4730)	
0.93 (4820)	0.89	(4730)	
0.85 (4620)	0.89	(4730)	
0.80 (4470)	0.70	(4180)	
0.78 (4420)	0.70	(4180)	
Σ = 45,710		Σ = 43,660	

What is the actual air flow rate and velocity? What is the standard flow rate?

$$V = 174\sqrt{P_v\frac{(t+460)}{K\times d}}, \text{ where:}$$

V = duct velocity, fpm
P_v = velocity pressure, inches of water
t = air stream temperature, °F
K = relative density for altitude (= 1.0 for altitudes < 1000 ft)
d = relative density correction factor for moisture. (Note that increased amounts of water vapor reduce the gas density since water vapor molecules weigh less than air molecules: 18 *vs.* 29.)

The moisture content for 140°F dew point air = 0.17 pound of water/pound of dry air. From standard psychrometric tables and charts, d = 0.918.

Taking the first velocity pressure reading of **0.70 in.** above:

$$V = 174\sqrt{(0.70)\frac{(300+460)}{(1)(0.918)}} = 4180 \text{ fpm. Repeat this for the other 19 readings!}$$

Therefore, $V_{avg} = \dfrac{45,710 \text{ fpm} + 43,660 \text{ fpm}}{20} = 4470 \text{ fpm}$

Area of six-ft diameter duct $= \dfrac{\pi}{4} \times (6 \text{ ft})^2 = 28.3 \text{ ft}^2$

Air flow rate $= 28.3 \text{ ft}^2 \times 4470 \text{ fpm} = 126{,}500$ actual cfm

Since the temperature is 300°F, the standard flow rate is:

$126{,}500 \text{ cfm} \times \dfrac{460 + 70}{460 + 300} = 88{,}200$ standard cfm

Answer: 126,500 actual cfm. 88,200 standard cfm. 4470 fpm duct velocity. If the gas was cooled to standard conditions, water condensation would occur.

422. A ventilation system branch duct with a 10,000 cfm design volume has a static pressure of –2.1 in. of water at the branch entry. The main, carrying a volume of 50,000 cfm, has a static pressure of –2.40 in. of water where the branch enters. What volume will be drawn through the branch at a balanced condition?

$Q_b = Q_o \sqrt{\dfrac{P_b}{P_o}}$, where:

Q_b = air flow volume required for balance

Q_o = design air flow volume

P_b = static pressure required for balance

P_o = static pressure originally calculated for balance

$Q_b = 10{,}000 \text{ cfm} \times \sqrt{\dfrac{2.40}{2.10}} = 10{,}700 \text{ cfm}$

Answer: 10,700 cubic feet of air/min.

423. An exhaust ventilation hood is required for gas tungsten-arc welding of non-ferrous metals at an assembly line production operation. The hood face can be no closer than nine in from the arc. A six-in flange can be installed on the 8 × 12 in. hood face. Because of variable cross-drafts which are difficult to control, a capture velocity of 300 fpm is desired. What exhaust volume is required?

$Q = K (10 \ x^2 + A) \ V_x$, where:

Q = exhaust volume, cfm

X = the distance from the center of the hood face to the farthest point of welding fume, smoke, and gases release, feet

A = hood face area (not including the flange), ft^2

V_x = minimum capture velocity, fpm

K = 1.0 for an unflanged hood; 0.75 for a large-flanged hood

$$Q = 0.75 \left[10\,(0.75\,\text{ft})^2 + \left(\frac{8\,\text{inches} \times 12\,\text{inches}}{144\,\text{in}^2/\text{ft}^2} \right) \right] 300\,\text{fpm} = 1416\,\text{cfm}$$

Answer: 1416 cubic ft of air/min. Be cognizant of the generation of ozone gas a considerable distance from the arc, such as occurs with TIG welding of aluminum.

424. In the preceding problem, select a round duct which will provide a minimum duct transport velocity of 3000 fpm.

$$A = \frac{Q}{V} = \frac{1416\,\text{cfm}}{3000\,\text{fpm}} = 0.472\,\text{ft}^2$$

A nine-in. internal diameter duct has a cross-sectional area of 0.4418 ft².

$$V = \frac{Q}{A} = \frac{1416\,\text{cfm}}{0.4418\,\text{fpm}} = 3205\,\text{fpm}$$

Answer: A nine-in. I.D. duct. When selecting a duct to achieve the minimum transport velocity, typically one should go to the next smaller size so that pollutant transport minimum velocity is achieved. For welding fume, in general, a minimum duct transport velocity is typically 2000 fpm. In some cases, 3000 fpm would be a better design choice.

425. Refer to problem 419. If there is 60% condensation of the water vapor in the 70°F and 90.3% relative humidity air, how much water (in gallons) would be condensed per hour from a single-pass ventilation system delivering 20,000 cfm?

vapor pressure of water at 70°F = 18.68 mm Hg

0.903 × 18.68 mm Hg = 16.87 mg Hg

$$\text{ppm}\,H_2O\,\text{vapor} = \frac{16.87\,\text{mm Hg}}{760\,\text{mm Hg}} \times 10^6 = 22{,}197\,\text{ppm}$$

$$\frac{\text{mg}\,H_2O}{M^3} = \frac{\text{ppm} \times \text{molecular weight}}{24.13\,\text{L/gram-mole}} = \frac{22{,}197\,\text{ppm} \times 18}{24.13} = \frac{16{,}558\,\text{mg}\,H_2O}{M^3}$$

$$16{,}558\,\text{mg}\,H_2O/M^3 = 468.9\,\text{mg/ft}^3$$

$$\frac{468.9\,\text{mg}}{\text{ft}^3} \times \frac{20{,}000\,\text{ft}^3}{\text{minute}} \times \frac{60\,\text{minutes}}{\text{hour}} = \frac{5.63 \times 10^8\,\text{mg}\,H_2O}{\text{hour}} =$$

$$\frac{1241\,\text{pounds}\,H_2O}{\text{hour}}$$

since 60% of the water vapor condenses: 0.6 (1241 lb/hr) = 744.6 lb/hr

$$\frac{744.6 \text{ lb/hour}}{8.33 \text{ lb/gallon}} = \frac{89.4 \text{ gallons}}{\text{hour}}$$

Answer: 89.4 gallons of water vapor will condense into liquid water per hr.

426. Community air was sampled for 30 days and 17 hrs at 7.3 cfm through an 8 × 10 in. high-efficiency particulate filter. A two cm × two cm section of this filter contained 89 micrograms of lead. What was the average airborne lead concentration during the sampling period (assuming lead mass was evenly distributed across the filter face)?

[30 days × (24 hrs/day)] + 17 hrs = 737 hrs = 44,220 min

44,220 min × 7.3 cfm = 322,806 ft^3 = 9140.8 M^3

4 cm^2 = 0.62 in^2 (80 in^2/0.62 in^2) = 129.03 = filter area multiplication factor

89 mcg Pb × 129.03 = 11,484 mcg Pb

$$\frac{11,484 \text{ mcg Pb}}{9140.8 \text{ M}^3} = \frac{1.26 \text{ mcg Pb}}{\text{M}^3}$$

Answer: 1.26 mcg Pb/M^3. Los Angeles air, for example, before organic lead was eliminated from gasoline, typically had an arithmetic mean and a geometric mean of 1.7 mcg Pb/M^3 and 1.08 mcg Pb/M^3, respectively.

427. A 120-liter cylinder of nitric oxide at 5.2 atmospheres and 20°C developed a leak. When the leak was repaired, 2.1 atmospheres of nitric oxide gas remained in the cylinder, which was still at 20°C. How many moles and grams of NO gas escaped?

moles of gas escaped = original moles of gas – final moles of gas =

$$\frac{(5.2 \text{ atm})(120 \text{ L})}{(0.0821)(293 \text{ K})} - \frac{(2.1 \text{ atm})(120 \text{ L})}{(0.0821)(293 \text{ K})} = 25.94 \text{ moles} - 10.48 \text{ moles} =$$

15.46 moles

15.46 mols NO × 30.01 g/mol = 463.95 g of NO

Answer: 15.46 mols and 463.95 g of nitric oxide gas escaped from the cylinder.

428. If, in the preceding problem, NO gas escaped into a 30 × 80 × 18 ft room without ventilation, what NO gas concentration can be predicted after thorough mixing?

15.46 mols NO × 30.01 g/mol = 463.95 g

$30 \times 80 \times 18$ ft $= 43{,}200$ ft$^3 = 1223.3$ M^3

$$\text{ppm} = \dfrac{\dfrac{463{,}950 \text{ mg}}{1223.3 \text{ M}^3} \times 24.45}{30.01} = 309 \text{ ppm NO}$$

Answer: 309 ppm NO (TLV = 25 ppm). NO converts into NO_2 at a steady rate following 1st order decay reaction kinetics, so the atmosphere in the room contains a mixture of noxious gases: $2 \text{ NO} + O_2 \rightarrow 2 \text{ NO}_2 \rightarrow N_2O_4$.

429. A chemical operator inadvertently poured 890 g of potassium iodide into a 70-gallon vat of sulfuric acid. There was excess sulfuric acid to oxidize the reducing agent (KI) to hydrogen sulfide. What volume of H_2S was produced and released to the atmosphere assuming negligible solubility of the H_2S in the sulfuric acid?

$8 \text{ KI} + \text{xs } 5 \text{ H}_2\text{SO}_4 \rightarrow 4 \text{ K}_2\text{SO}_4 + 4 \text{ I}_2\uparrow + \text{H}_2\text{S}\uparrow + 4 \text{ H}_2\text{O}$

Eight mols of KI generate one mol of H_2S, or:

$$(890 \text{ g KI}) \left[\frac{1 \text{ mole KI}}{166 \text{ g KI}} \right]\left[\frac{1 \text{ mole H}_2\text{S}}{8 \text{ moles KI}} \right]\left[\frac{24.45 \text{ L at NTP}}{\text{mole H}_2\text{S}} \right] = 16.39 \text{ liters of H}_2\text{S}.$$

Answer: 16.39 L H_2S. 16,390,000 L of air would be necessary to dilute this rotten egg odor gas to one ppm. Iodine vapor would also be generated.

430. Calculate the lower and upper confidence limits for a time-weighted air sampling result of 28 ppm for a worker. The PEL/TLV for this organic vapor is 25 ppm. The combined analytical, sampling, and other cumulative errors are $\pm 19.5\%$ (sampling analytical error, SAE; See problem 5 for the procedure to calculate SAE).

$$\text{LCL} = \frac{\text{EC}}{\text{PEL or TLV}} - \text{SAE} \quad \text{and UCL} = \frac{\text{EC}}{\text{PEL or TLV}} + \text{SAE, where:}$$

LCL and UCL = lower and upper confidence limits, respectively. The UCL is rarely used in normal industrial hygiene compliance determinations.

EC = exposure concentration (ppm, f/cc, or mg/M^3)

PEL or TLV = permissible exposure limit or threshold limit value

SAE = sampling and analytical error as a decimal (absolute — disregard algebraic signs)

LCL = $(28/25) - 0.195 = 0.925$

UCL = $(28/25) + 0.195 = 1.315$

Answer: Lower confidence limit = 0.925. Upper confidence limit = 1.315. Since the LCL is less than one, and assuming a normal distribution for exposure levels, a violation of the PEL and TLV has not occurred. However, since the exposure is well above the action level, industrial hygiene controls are warranted. The level of confidence in this statistical parameter is 95% which means that, to a 95% degree of certainty, the true value lies between 92.5% and 131.5% of the reported value, or 25.9 ppm to 36.8 ppm.

431. Automobiles are multi-coat spray painted with an average of 1.1 gallons of high-solids paint (43% solids) per vehicle. The paint solvents are primarily a mixture of aromatic hydrocarbons, ketones, alcohols, and naphtha. The automobile bodies are baked to dryness in an oven at 350°F for 13 min. An air pollution control engineer wants to know what the solvent evaporation rates are in pints per vehicle and the total solvent vapor emissions if this assembly plant produces 42 vehicles per hr. Can you help him?

Essentially 57% of the paint mass evaporates in the baking oven. A small amount evaporates in the vestibules and spray booths leading to the oven. Regardless, all contribute to volatile organic carbon atmospheric emissions (VOCs).

(0.57) (1.1 gallons/vehicle) (eight pints/gallon) = 5 pints/vehicle

5 pints/vehicle × 42 vehicles/hr = 210 pints/hr = 3.5 pints/min

Answers: 5 pints/vehicle. 3.5 pints/min. 210 pints/hr. Emissions in baking ovens lend themselves to engineering controls, whereas vapors released outside of the ovens are fugitive emissions which are more difficult to control.

432. How much water vapor is contained in a building 15 meters × 30 meters × 5 meters if the relative humidity is 40% and the dry bulb temperature is 86°F (30°C)? The vapor pressure of water at this temperature is 31.82 torr (31.82 mm Hg).

15 m × 30 m × 5 m = 2250 M^3 = 2,250,000 L

$P = (0.40)$ $(P_{vapor}) = (0.40)$ (31.82 torr) = 12.73 torr (partial pressure due to water vapor)

$$n = \frac{PV}{RT} = \frac{\left[(12.73/760)\text{atm}\right](2,250,000 \text{ L})}{(0.0821 \text{ L-atm/mole-K})(303.15 \text{ K})} = 1514.2 \text{ moles of water vapor}$$

$$(1514.2 \text{ moles})\left[\frac{18.0 \text{ g}}{\text{mole}}\right]\left[\frac{1 \text{ kg}}{10^3 \text{ g}}\right] = 2726 \text{ kg of water vapor}$$

Answer: 27.26 kilograms of water vapor (60.1 pounds).

433. Confined, occupied spaces without outside fresh air ventilation (e.g., subma-
rines, space capsules, diving bells) can accumulate significant levels of carbon
dioxide gas from human respiration. Lithium oxide, Li_2O, is the most efficient
CO_2 gas scavenger. What is the CO_2 absorption efficiency of Li_2O in liters of
CO_2 (at STP) per kilogram?

$$Li_2O + CO_2 \rightarrow Li_2CO_3 \text{ (lithium carbonate)}$$

$$\left(1 \text{ kg } Li_2O\right)\left[\frac{100 \text{ g}}{\text{kg}}\right]\left[\frac{1 \text{ mole } Li_2O}{29.88 \text{ g } Li_2O}\right]\left[\frac{22.4 \text{ L at STP}}{\text{mole of } CO_2}\right]\left[\frac{1 \text{ mole } CO_2}{1 \text{ mole } Li_2O}\right]$$

$$= 749.7 \text{ liters}$$

Answer: Each kilogram of lithium oxide will theoretically absorb 749.7 L of
carbon dioxide gas at STP. Knowing the maximum CO_2 gas generation rate
for the number of people present (body mass, metabolic rates, etc.), the amount
of lithium oxide required for an air cleansing system can be calculated. At
STP, CO_2 weighs 1977 $g/M^3 = 1.977$ g/L. Exhaled air contains up to 5.6%
CO_2 by volume.

434. A brine-ammonia refrigeration plant loses an average of 35 ft^3 of ammonia
every day from leaking valves and process equipment as determined from the
purchase records. What is the dilution ventilation air volume necessary to
control to no more than 10% of the TLV of 25 ppm for NH_3?

$$\frac{35 \text{ ft}^3/\text{day}}{24 \text{ hours/day}} = \frac{1.458 \text{ ft}^3}{\text{hour}} \times \frac{\text{hour}}{60 \text{ minutes}} = 0.0243 \text{ ft}^3/\text{minute}$$

$$25 \text{ ppm} \times 0.1 = 2.5 \text{ ppm}$$

$$\frac{0.0243 \text{ ft}^3/\text{minute}}{2.5 \text{ ppm}} \times 10^6 = 9720 \text{ cfm}$$

Answer: 9720 cubic feet of fresh dilution air/min assuming good mixing.
Repair the leaking equipment first; then consider increasing the dilution ventilation.

435. An elderly, senile man, in an attempt to cool his top-floor, non-airconditioned
and nonventilated room, placed a 50-pound block of dry ice next to his bed.
When his body was recovered from the 12 $\times$ 16 $\times$ 8 ft room, the dry ice
block weighed about 22 pounds. What was the most likely average oxygen
concentration in the air in the room just before the man's body was discovered?
What was the most likely average CO_2 gas concentration?

Assume a worst case scenario in which there was no ventilation or air exchange
in the room.

$$\frac{50 \text{ lb } CO_2 - 22 \text{ lb } CO_2}{12 \text{ ft} \times 16 \text{ ft} \times 8 \text{ ft}} = \frac{28 \text{ lb}}{1536 \text{ ft}^3} = \frac{0.0182 \text{ lb } CO_2}{ft^3}$$

$$\frac{0.0182 \text{ lb}}{ft^3} \times \frac{35.315 \text{ ft}^3}{M^3} \times \frac{kg}{2.205 \text{ lb}} \times \frac{1000 \text{ g}}{kg} \times \frac{1000 \text{ mg}}{g} = \frac{291,489 \text{ mg } CO_2}{M^3}$$

$$ppm = \frac{\dfrac{291,489 \text{ mg}}{M^3} \times 24.45}{44.01} = 161,938 \text{ ppm } CO_2 = 16.19\% \ CO_2$$

100% air $-$ 16.19% CO_2 = 83.81% air remaining

83.81% air $\times$ 0.2095 = 17.55% oxygen

Answer: About 17.6% oxygen and 16.2% CO_2. OSHA regards an atmosphere containing less than 19.5% oxygen as oxygen-deficient. The NIOSH IDLH for CO_2 is 40,000 ppm (4%). Protracted exposure of healthy adults to air containing 17.5% oxygen, while physiologically taxing, would probably not cause significant adverse effects. In this case, however, CO_2 contributes toxic effects as a respiratory and cardiac stimulant, prolonged CO_2 inhalation can produce metabolic acidosis, the gentleman was old (with, possibly, poor cardiovascular status), and heat strain was, most likely, an added myocardial stressor. Moreover, since the block of dry ice was located next to the bed where his body was found, pockets of carbon dioxide gas (e.g., say, 50%?) could have surrounded his head for times sufficiently long to produce anoxia, asphyxiation, and a heart attack.

436. The carbon dioxide gas exhaled by laboratory rats into the test chambers used in experimental toxicology will be absorbed so that the chamber atmosphere can be recirculated. How much carbon dioxide gas can be absorbed into a solution which contains 900 g of potassium hydroxide? Assume 100% CO_2 gas absorption. The molecular weights of potassium hydroxide and carbon dioxide are 56.11 and 44.01, respectively.

CO_2 + 2 KOH $\rightarrow$ K_2CO_3 + H_2O (K_2CO_3, by the way, is also known as potash — originally obtained from the potassium-rich ashes of wood fires. Potash and lard, or animal fat, were mixed to produce the original soap.)

Therefore, two mols of KOH (or NaOH) will absorb one mol of CO_2 gas.

$$\frac{900 \text{ g KOH}}{56.11 \text{ g/mole}} = 16.04 \text{ moles of KOH}$$

16.04 mols of KOH will absorb 8.02 mols of CO_2 gas.

$$8.02 \text{ moles } CO_2 \times \frac{44.01 \text{ grams}}{mole} = 352.96 \text{ grams of } CO_2$$

Answer: 353 g of CO_2. Refer to problem 433 and see how lithium oxide generally is more efficient (kilogram per kilogram) as a CO_2 gas scavenger. The higher cost of lithium oxide, however, might argue for the purchase of the less expensive KOH or NaOH.

437. What is the density of a mixture of 500 ppm methyl alcohol vapor in air?

The "apparent" molecular weight of dry air (based on the volume % composition of nitrogen, oxygen, argon, CO_2, etc.) is nearly 29 (exactly = 28.941 g/g-mol of air). The density of dry air is 1.2 mg/mL at 25°C and 760 mm Hg. The density of air is defined as 1.00 (no units, a relative number). Since methanol has a molecular weight of 32.01, the density of 100% MeOH vapor (i.e., 10^6 ppm MeOH) compared to air is (since a mole of MeOH has greater mass than a "mole" of air):

$$\frac{32.01}{28.941} = 1.106 \qquad 500 \text{ ppm MeOH} = 0.05\%$$

for MeOH vapor: $0.0005 \times 1.106 = 0.00055$
for air: $\underline{0.9995 \times 1.00} = \underline{0.9995}$
total $\quad 1.0000 \qquad\qquad 1.00005$

Answer: 1.00005 (no units). This is essentially the same as air since (1) MeOH has a molecular weight close to that of air, and (2) 500 ppm MeOH is a relatively dilute concentration of vapor, i.e., 50/1,000,000.

438. The U.S. Nuclear Regulatory Commission established in their document NUREG-1391 (*Chemical Toxicity of Uranium Hexafluoride Compared to Acute Effects of Radiation*) the values for uranium and hydrogen fluoride gas that should be used for accident analysis purposes at facilities that process large quantities of UF_6. UF_6 reacts exothermically with moisture in the air to form uranyl fluoride (UO_2F_2) and hydrogen fluoride (HF):

$$UF_6 + 2 H_2O \rightarrow UO_2F_2 + 4 HF + heat$$

The human inhalation exposure value for hydrogen fluoride gas is:

$$HF \text{ concentration} = \left[\frac{25 \text{ mg}}{M^3}\right]\sqrt{\frac{30 \text{ minutes}}{t}}, \text{ where t is the time in minutes of}$$

the duration of exposure. If the exposure duration is 15 min, what is the maximum acceptable HF gas concentration in ppm using this NRC criteria?

$$HF \text{ concentration} = \left[\frac{25 \text{ mg}}{M^3}\right]\sqrt{\frac{30 \text{ minutes}}{15 \text{ minutes}}} = \frac{35.4 \text{ mg}}{M^3}$$

$$\text{ppm} = \frac{\frac{mg}{M^3} \times 24.45}{\text{molecular weight of HF}} = \frac{\frac{35.4 \, mg}{M^3} \times 24.45}{20.01} = 43.3 \text{ ppm HF}$$

Answer: Approximately 43 ppm HF. This exceeds the ACGIH "ceiling" TLV of three ppm. The AIHA Emergency Response Planning Guideline for HF gas for up to one-hr exposure is two ppm (without most people "experiencing other than mild transient adverse health effects or perceiving a clearly defined objectionable odor"); for an ERPG-2 ("nearly all individuals could be exposed for up to one hr without experiencing or developing irreversible or other serious health effects or symptoms that could impair their abilities to take protective action"), the value for HF is 20 ppm; for an ERPG-3 ("nearly all individuals could be exposed for up to one hr without experiencing life-threatening health effects"), the value for HF gas is 50 ppm.

439. If the air temperature (dry bulb) is 86°F, and the dew point is 68°F, what is the relative humidity? The pressure of saturated water vapor at 86°F is 31.82 mm Hg. At 68°F, the pressure of saturated water vapor is 17.54 mm Hg.

$$\text{relative humidity} = \frac{\text{saturated pressure of water vapor in air dew point}}{\text{pressure of water vapor in saturated air at } 86°F} =$$

$$\frac{17.54 \text{ mm Hg}}{31.82 \text{ mm Hg}} \times 100 = 55.1\%$$

Answer: 55.1% relative humidity. The dew point is the temperature at which the atmosphere would be saturated with the contained water vapor.

440. Fifty pounds of sodium hydroxide pellets are rapidly dumped into an open top steel tank containing 40 gallons of water. The solution becomes very hot, and there is evolution of an irritating caustic mist aerosol. After returning to room temperature, 42.9 gallons of solution remain in the tank. After correcting for the blank sodium content, the solution contained 80,070 mg Na/L. If the building in which the NaOH aerosol was released was 35 × 120 × 18 ft, how much sodium hydroxide was released into the air? What was the average mist concentration if the room did not have ventilation?

Molecular weights of Na and NaOH, respectively, are 23 and 40.

$$42.9 \text{ gallons} \times \frac{80.07 \text{ g Na}}{L} \times \frac{1 \, L}{0.264 \text{ gallon}} \times \frac{40}{23} \times \frac{1 \, lb}{453.59 \, g} = 49.88 \text{ lb NaOH}$$

$$\frac{(50 \text{ lb NaOH} - 49.88 \text{ lb NaOH})}{35 \text{ ft} \times 120 \text{ ft} \times 18 \text{ ft}} = \frac{0.12 \text{ lb NaOH}}{75,600 \text{ ft}^3}$$

$$\frac{0.12 \text{ lb NaOH}}{75,600 \text{ ft}^3} \times \frac{35.314 \text{ ft}^3}{M^3} \times \frac{459.59 \text{ g}}{\text{lb}} \times \frac{1000 \text{ mg}}{\text{g}} = \frac{25.8 \text{ mg NaOH}}{M^3}$$

Answer: 0.12 lb of NaOH was released into the air. The average concentration of NaOH mist in the air was 25.8 mg/M^3 (ACGIH ceiling TLV = two mg/M^3). The mist concentration near the tank, obviously, initially would have been much higher.

441. A grave explosion hazard presents in the molten materials industry when materials above 100°C are charged in vessels which contain liquid water. Huge explosions resulting in fatalities, injuries, and substantial property damage have occurred when, for example, molten metal is poured into receiving vessels, or onto a floor, where water has accumulated. The higher the molten material temperature, the greater the amount of water, and the higher the rate of addition all contributively add to the explosion's magnitude. Calculate the volume of steam produced when one ton of molten iron is carelessly added to a four-ft internal diameter pouring ladle containing water three in deep. The atmospheric pressure is 720 mm Hg.

water volume = $\pi r^2 h = \pi (2 \text{ ft})^2 \times 0.25 \text{ ft} = 3.1416 \text{ ft}^3 = 88,960$ mL of liquid water

density of water = 1.00 g/mL   PV = nRT

$$\frac{88960 \text{ g H}_2\text{O}}{18.0 \text{ g/mole}} = 4942 \text{ moles of water} \qquad T = 100°C + 273 \text{ K} = 373 \text{ K}$$

$$V = \frac{nRT}{P} = \frac{(4942 \text{ moles})(0.0821 \text{ atm-liter/mole-K})(373 \text{ K})}{\dfrac{720 \text{ mm Hg}}{760 \text{ mm Hg}} = 0.947 \text{ atmosphere}} = 159,810 \text{ L}$$

Answer: The liquid volume of 88.96 L of water explosively expands into a 159,810 L steam cloud. For this reason, torches or other means are applied to keep the interiors of receiving vessels bone dry before addition of molten metals, glass, plastics, etc. Besides the plume of steam, molten metal spews from the ladle.

442. A reasonable, reliable microscopic quantification of fibers on air filters using phase contrast microscopy (PCM) is 10 fibers per 100 fields. What are the equivalent, reliable PCM quantifications for sampling volumes of 3000, 5000, and 7500 L of air?

855 mm² is the collection area of a 37 mm diameter air filter. Obviously, a smaller diameter filter (e.g., 25 mm) could improve detection limits.

0.003 mm² is the size of a typical field of view for a PC microscope, which varies between 0.003 mm² and 0.006 mm² for different microscopes. Larger fields of view will improve (decrease) the limit of reliable quantification.

$$\text{Quantification limit} = \frac{10 \text{ fibers/100 fields}}{3000 \text{ liters}} \times \frac{855 \text{ mm}^2}{0.003 \text{ mm}^2} \times \frac{1 \text{ liter}}{1000 \text{ cc}} = \frac{0.01 \text{ fiber}}{\text{cc}}$$

$$\frac{0.01 \text{ f/cc}}{\text{xf/cc}} = \frac{5000 \text{ L}}{3000 \text{ L}} \quad x = 0.006 \text{ f/cc} \qquad \frac{0.01 \text{ f/cc}}{\text{xf/cc}} = \frac{7500 \text{ L}}{3000 \text{ L}} \quad x = 0.004 \text{ f/cc}$$

Answers: 0.01 f/cc, 0.006 f/cc, and 0.004 f/cc, respectively for air sampling volumes of 3000, 5000, and 7500 L. NIOSH PCM Method 740 will improve the reliable limit of quantification for the same air sampling volumes.

443. Johnny yanks the thermometer from his mouth and throws it against the wall. The thermometer smashes releasing 0.5 mL of mercury which spills upon the carpet of his bedroom. Johnny lives in a 1500 square foot bungalow without a basement. The ceilings are eight-ft high. Assuming absolutely no ventilation of his house, what maximum average mercury vapor concentration can be, in time, achieved within this house? Assume no adsorption on surfaces.

This problem can be approached in no less than three ways: collect air samples until the equilibrium Hg vapor concentrations are reached; perform calculations assuming all mercury evaporates into the house atmosphere; and calculate the maximum vapor concentration based upon mercury's vapor pressure. Mercury's molecular weight, density, and vapor pressure at normal temperature are 200.6, 13.6 g/mL, and 0.0012 mm Hg, respectively.

mass/volume method:

0.5 mL × 13.6 g/mL = 6.8 g = 6800 mg of mercury evaporate
1500 ft² × 8 ft = 12,000 ft³ = 339.8 M³

$$\frac{6800 \text{ mg}}{339.8 \text{ M}^3} = 20.01 \text{ mg/M}^3$$

vapor pressure method:

$$\frac{0.0012 \text{ mm Hg}}{760 \text{ mm Hg}} \times 10^6 = 1.579 \text{ ppm Hg vapor at saturation}$$

$$\text{mg Hg/M}^3 = \frac{\text{ppm} \times \text{mol. wt.}}{24.45 \text{ L/g-mole}} = \frac{1.579 \text{ ppm} \times 200.6}{24.45} = 12.95 \text{ mg/M}^3$$

Answer: Note the large variance between the calculated concentrations. Since air cannot contain more than saturated concentrations, the correct answer is 12.95 mg/M³. This is considerably greater than ACGIH's threshold limit value for mercury of 0.025 mg/M³ for work places which, in turn, is greater than that which should be present inside a residence. Without recovery of the spilled mercury, encapsulation or chemical binding of the mercury, and/or good ventilation, Johnny risks mercury poisoning. In summary, there is a sufficient amount of liquid mercury in a standard household thermometer to saturate air in a 1500 square foot house with poisonous vapors.

444. Estimate the volumetric air flow rate above a slab of recently rolled steel which is 30 ft long and eight ft wide. The estimated rate of heat loss from the 10-in. thick slab is an average of 5,500,000 BTU/hr for the first four hrs out of the rolling mill. Assume the slab is surrounded on four sides by 40-ft high aluminum sheet radiation panels so that most of the radiant portion of heat loss is converted into sensible heat and hot air convection currents. The height of the steel rolling mill is 100 ft above the floor.

The air motion may be estimated by: $Q_z = 1.9\, Z^{1.5}\, \sqrt[3]{q}$, where:

Q_z = air flow rate at effective height Z (in feet), in cfm. The Z factor accounts for an envelope of air which expands as it rises from the hot body

q = convection heat loss from the hot body, BTU/hr

$Z = Y + 2\,B$

Y = actual height above hot body, feet (in this case, at the top of the heat shields)

B = largest horizontal dimension of hot body, feet (Note how the shorter dimension of width does not enter into the calculations.)

$Z = 40\text{ ft} + (2 \times 30\text{ ft}) = 100\text{ ft}$ (coincident with the height of the steel rolling mill)

$$Q = (1.9)\,(100)^{1.5}\left[\sqrt[3]{5,500,00}\,\right] = 335,000\text{ cfm}$$

Answer: 335,000 cfm.

445. Assume a branch in an exhaust ventilation system has a design volume of 10,000 cfm with a calculated static pressure, P_s, of 2.10 in. water gauge. The main exhaust duct, carrying an exhaust volume of 50,000 cfm, has a P_s of 2.40 in. water gauge where the branch enters. What volume of air will be drawn through the branch at balanced conditions?

$$Q_b = 10,000\text{ cfm}\sqrt{\frac{2.40\text{ inches}}{2.10\text{ inches}}} = 10.700\text{ cfm}$$

Answer: 10,700 cubic feet of air/min.

446. What is the worst-case peak blood methyl alcohol level in a 70-kg man exposed to the ACGIH TLV of 200 ppm?

"Standard 70-kg man" has a 60% water content and inhales 10 cubic meters of air during an eight-hr workday.

$$\frac{mg}{M^3} = \frac{ppm \times molecular\ weight}{24.45} = \frac{200 \times 32.04}{24.45} = 262\ mg/M^3$$

262 mg/M³ × 10 M³ = 2620 mg MeOH   70 kg × 0.6 = 42 kg H_2O = 42 L H_2O

peak blood level of MeOH = 2620 mg/42 L = 62 mg/L = 6.2 mg/dL

Answer: 6.2 mg/dL. This is close to 5% of the dose reported to cause acute irreversible toxic effects. Furthermore, since the blood half-life of methanol is three hrs, the body burden would be negligible when the worker returned to his job the next day. It appears highly unlikely that this dose could cause ocular toxicity. However, MeOH is absorbed through intact healthy skin, and industrial hygienists, workers, and the workers' supervisors must be mindful of this potential route of exposure. The preceding calculations assume that most of the MeOH is in the aqeous compartment of the body; some dissolves in the lipid compartment as well.

447. How many tons of air are exhausted yearly from a laboratory exhaust hood with full open face dimensions of 24 × 48 in.? The face capture velocities are 130, 140, 120, 120, 140, 130, 110, 120, and 110 cfm/ft². The hood operates eight hrs per day, five days a week, for 50 weeks a year.

$$\left[\frac{130+140+120+120+140+130+110+120+110}{9}\right]cfm/ft^2 = 124.4\ cfm/ft^2$$

24 × 48 in. = 2 × 4 ft = 8 ft²   8 ft² × 124.4 cfm/ft² = 995.2 cfm

$$\frac{995.2\ ft^3}{minute} \times \frac{60\ minutes}{hour} \times \frac{8\ hours}{day} \times \frac{5\ days}{week} \times \frac{50\ weeks}{year} \times \frac{0.075\ lb}{ft^3} \times$$

$$\frac{ton}{2000\ lb} = 4478.4\ tons/year$$

Answer: 4478.4 tons of air are exhausted each year.

448. What is the solvent vapor emission rate from an open surface tank (2.5 × 4 ft) containing trichloroethylene at room temperature? The molecular weight and the vapor pressure of TCE are 131.4 g/g-mol and 58 mm Hg, respectively. The barometric pressure is 748 mm Hg. The air velocity passing over the surface of the tank is 200 ft/min (fpm). The air temperature is 78°F.

The following formula can be used to estimate the emission rate, q, in grams per second:

$$Q = \frac{8.24 \times 10^{-8} \times M^{0.835} \times P\left[\dfrac{1}{29} + \dfrac{1}{M}\right]^{0.25} \times U^{0.5} \times A}{T^{0.05} \times L^{0.5} \times P_t^{0.5}}$$

M is the molecular weight. P is the vapor pressure, mm Hg. U is the air velocity, fpm. A is the surface area, cm². T is the air temperature, kelvin. L is the length of the liquid surface, cm. Pt is the total pressure, atmospheres.

A = 2.5 × 4 ft = 10.0 ft² = 9290.3 cm²   T = 78°F = 298.7 kelvin
L = 4 ft = 121.9 cm P_t = (748 mm Hg/760 mm Hg) = 0.984 atm

$$q = \frac{8.24 \times 10^{-8} \times 131.4^{0.835} \times 58\left[\dfrac{1}{29} + \dfrac{1}{131.4}\right]^{0.25} \times 200^{0.5} \times 9290.3}{298.7^{0.05} \times 121.9^{0.5} \times 0.984^{0.5}} = 1.15 \text{ g/sec}$$

Answer: 1.15 g of trichloroethylene evaporate every second. Refer to problem 449.

449. Referring to problem 448, the depth of the trichloroethylene in the tank was 23 1/2 in. on Monday at 8:00 a.m. On Friday, at 4:00 p.m., the solvent depth was 18 1/2 in.. If the system was not used for de-greasing (That is, there was no physical carry-out of the liquid solvent, and there were no additions), what was the vapor emission rate assuming system values identical to those specified in problem 448? Assume that the exhaust ventilation system for the tank operated eight hrs per day. The specific gravity of trichloroethylene is 1.46 g/mL.

23.5 in. – 18.5 in. = 5 in. evaporative loss   2.5 ft = 30 in.   4 ft = 48 in.
5 × 30 × 48 in. = 7200 in³ = 117,987 cM³
117,987 mL × 1.46 g/mL = 172,261 g
8 hrs/day × 5 days × 60 min/hr × 60 sec/min = 144,000 sec
172,261 g/144,000 sec = 1.20 g/sec

Answer: 1.20 g of trichloroethylene evaporated every second. The semi-empirical approach used in this problem provides a result essentially identical to the calculated theoretical amount applying the equation in problem 448.

450. Pure helium gas passes through a rotameter at an indicated flow rate of 1.7 L/min. The rotameter was calibrated at 29.9 in. of water and 70°F. Helium has a specific gravity of 0.138 at 70°F and 29.9 in. of water (air = 1.00). If the helium temperature is 100°F, and its pressure is 33 in. of water, what is the actual helium gas flow rate?

$$k = \sqrt{\frac{460 + 100}{460 + 70} \times \frac{29.9}{33} \times \frac{1.00}{0.138}} = 2.63 \qquad 1.7 \text{ L/m} \times 2.63 = 4.47 \text{ L/m}$$

Answer: 4.47 L of helium per min, or 2.63 times greater than the rate indicated on the rotameter.

Index to Problems